Construction Contract Claims

Macmillan Building and Surveying Series
Series Editor: IVOR H. SEELEY
Emeritus Professor, The Nottingham Trent University

Advanced Building Measurement, second edition Ivor H. Seeley
Advanced Valuation Diane Butler and David Richmond
An Introduction to Building Services Christopher A. Howard
Applied Valuation Diane Butler
Asset Valuation Michael Rayner
Building Economics, third edition Ivor H. Seeley
Building Maintenance, second edition Ivor H. Seeley
Building Maintenance Technology Lee How Son and George C. S. Yuen
Building Procurement Alan E. Turner
Building Project Appraisal Keith Hutchinson
Building Quantities Explained, fourth edition Ivor H. Seeley
Building Surveys, Reports and Dilapidations Ivor H. Seeley
Building Technology, fourth edition Ivor H. Seeley
Civil Engineering Contract Administration and Control, second edition Ivor H. Seeley
Civil Engineering Quantities, fifth edition Ivor H. Seeley
Civil Engineering Specification, second edition Ivor H. Seeley
Commercial Lease Renewals Philip Freedman and Eric F. Shapiro
Computers and Quantity Surveyors Adrian Smith
Construction Contract Claims Reg Thomas
Contract Planning and Contractual Procedures, third edition B. Cooke
Contract Planning Case Studies B. Cooke
Cost Estimation of Structures in Commercial Building Surinder Singh
Design-Build Explained David E. L. Janssens
Development Site Evaluation N. P. Taylor
Environmental Science in Building, third edition R. McMullan
Greener Buildings Stuart Johnson
Housing Associations Helen Cope
Housing Management: Changing Practice Christine Davies (*editor*)
Information and Technology Applications in Commercial Property Rosemary Feenan and Tim Dixon (*editors*)
Introduction to Valuation D. Richmond
Marketing and Property People Owen Bevan
Principles of Property Investment and Pricing, second edition W. D. Fraser
Property Valuation Techniques David Isaac and Terry Steley
Public Works Engineering Ivor H. Seeley
Quality Assurance in Building Alan Griffith
Quantity Surveying Practice Ivor H. Seeley
Recreation Planning and Development Neil Ravenscroft
Resource Management for Construction M. R. Canter
Small Building Works Management Alan Griffith
Structural Detailing, second edition P. Newton
Urban Land Economics and Public Policy, fourth edition P. N. Balchin, J. L. Kieve and G. H. Bull
Urban Renewal – Theory and Practice Chris Couch
1980 JCT Standard Form of Building Contract, second edition R. F. Fellows

Construction Contract Claims

Reg Thomas

BSc (Hons), FCIOB, ACIArb, MBIM

Director Overseas Development,

James R. Knowles

First published 1993 by
THE MACMILLAN PRESS LTD
Houndmills, Basingstoke, Hampshire RG21 2XS
and London
Companies and representatives
throughout the world

ISBN 0-333-55498-1 hardcover
ISBN 0-333-55499-X paperback

A catalogue record for this book is available from the British Library.

Reprinted 1994

Printed in Hong Kong

To my wife Joan

Contents

Foreword

The preparation and negotiation of claims has become an industry within an industry. In fact, during a period of recession it is one of the few sections of the construction industry which flourishes. It is not surprising therefore to see the publication of another book which deals with claims. There are a number of books on the market to do with claims but Reg Thomas's *Construction Contract Claims* has a number of features which are not very well catered for by the others.

The section dealing with claims prevention should be studied particularly by architects and engineers. Reg Thomas draws attention to the oft-adopted policy of assuming that the issue of information to contractors can be delayed with impunity on the grounds that the contractor himself is already in delay. The book argues that the contractor, in support of an application for an extension of time or a claim that time has become at large, may argue that even though he is in delay, completion to time would in any event have been impossible due to the late issue of information.

Claims settlement invariably becomes protracted and difficult where records are poor or non-existent. Great assistance is provided by the book with regard to the type of records which should be kept.

Most books dealing with construction law contain numerous interesting and relevant cases. This book is no exception. An advantage which this book has to offer is that as many construction cases have been brought before the courts in the last few years they are all included. A case which is likely to have a long-lasting effect upon the way in which claims are prepared and presented is *Wharf Properties and Another* v. *Eric Cumine Associates and Another* (1988). This case has thrown doubt on the preparation of global rolled-up claims and is dealt with in the book.

A criticism I levy against many books dealing with construction law is that they answer all the simple questions but studiously avoid those which are thorny. Reg Thomas seems to have developed his theme by highlighting the difficult contractual problems and providing cogent answers. In particular I like the sections dealing with concurrent delays and the contractual effect of variations issued after the contract completion date but before the date of practical completion.

The recovery of head office overheads is comprehensively dealt with in the book and an interesting aspect is reference to and explanation of the Eichleay formula used in the USA.

Whether a book is read or not is often dependent upon the style in which it is written. Some books are heavy going from the first page. Reg Thomas's *Construction Contract Claims* is written with a light touch and is easy to read, understand and digest and I have no hesitation in recommending it to all involved in the construction process, whether building, civils or engineering services.

Roger Knowles
FRICS FCIArb, Barrister

Preface

There are a number of excellent text books on construction law, contracts and claims. The author has referred to *Hudson's Building and Engineering Contracts, tenth edition* for a number of early cases, and readers are advised to refer to this invaluable source for a better understanding of many issues discussed in this book. Publications by James R. Knowles listed in the bibliography have also been invaluable in the preparation of this book and are recommended for further reading. Knowles' publications and summaries of the cases cited in References may be purchased from Knowles Publications, Wardle House, King Street, Knutsford, Cheshire WA16 6PD. The contents of this book are intended to present to readers a general view of the practical problems which exist and how they might be avoided or resolved. The views expressed by the author represent several years' experience of looking backwards at projects which have gone wrong. In practice, many projects go well, are completed without major claims, and where they do occur, they are often settled promptly, professionally and amicably. Unfortunately, there is an increasing incidence of claims, most of which are brought about by financial pressures which stretch the resources of consultants, contractors and subcontractors alike. Many firms do not have sufficient allowances built into their fees, or into the contract price, to carry out their obligations properly. Some firms lack sufficient staff with the skills required to manage projects efficiently and to deal with claims in a professional manner. Insufficient attention to training staff, so that they can be better prepared to deal with claims, is another reason for many of the problems which exist in the industry. Whilst many claims are well presented and dealt with professionally by the recipient, some of these failures are evidenced in the presentation and quality of some claims submitted by large and small firms alike and in the response made by some architects, engineers and quantity surveyors.

The chapters which follow attempt to guide readers through the history of developments in law and contracts so that they may understand more fully the reasons for good contracts administration as a means of avoiding or minimising the effects of claims for delay and disruption.

Some of the arguments and methods of quantifying claims in this book should be regarded as possible means of persuasion according to the circumstances and records which are available to support a claim. In some cases, a lack of records may not be fatal to a claim, but it may be an uphill battle to persuade the recipient of a claim to pay out large sums of money on the basis of hypothetical calculations which have no real foundation. Readers should be aware that there is no real substitute for good records when it comes to quantifying a claim for an extension of time or for additional payment. Nevertheless, if the contractor has been delayed at almost every turn, it must be right that he receives some relief and compensation so far as it can be established by applying commonsense according to the circumstances. As a consultant to contractors and subcontractors, a duty is owed to them to use every means available, providing that they are honest and justifiable, to obtain the best possible settlement of their claims. As a consultant to employers (or to contractors defending a claim from subcontractors), a duty is owed to them to defend all claims and to discredit any unmeritorious claims. Nevertheless, employers (and contractors as the case may be) will need to be advised on the possible worth of a claim in order to facilitate a decision as to settlement or arbitration or litigation.

Whilst some practitioners may seek refuge in cases in which claims have been rejected on the grounds that the records and/or the method of quantification were lacking, the author supports the view expressed in *Penvidic Contracting Co. Ltd* v. *International Nickel Co. of Canada Ltd* (1975) 53 DLR (3d) 748 (Quoting Davies J. in *Wood* v. *Grand Valley Railway Co*) – see *A Building Contract Casebook* by Dr Vincent Powell Smith and Michael Furmston at page 316.

> 'It was clearly impossible under the fact of that case to estimate with anything approaching to mathematical accuracy the damages sustained by the plaintiffs, but it seems to me clearly laid down there by the learned Judges that such an impossibility cannot "relieve the wrongdoer of the necessity of paying damages for his breach of contract" and that on the other hand the tribunal to estimate them, whether jury or Judge, must under such circumstances do "the best it can" and its conclusion will not be set aside even if '*the amount of the verdict is a matter of guess work.*' (emphasis added).

However, the above quotation should not be relied upon to cure all ills. The terms of the contract and other circumstances may require a more robust approach when defending any claim which is clearly deficient in the essential ingredients to justify anything less than total or partial rejection.

It is hoped that this book will provide useful guidance for those responsible for dealing with claims so that they can be resolved with the minimum cost and without any party being seriously disadvantaged.

Reginald W. Thomas
Spring 1992

Acknowledgements

The author expresses his sincere thanks to Roger Knowles for giving his consent to use of the extensive computer library facility of James R. Knowles, including notes and diagrams used for seminars conducted by the company, and for writing the Foreword to this book.

Particular mention and thanks must be given to Ann Glacki, head of James R. Knowles' library and author of BLISS (Building Law Information Subscriber Service) for her co-operation and assistance in searching for suitable cases and other reference material which have been invaluable for the preparation of this book. I also thank Peter Nuttall, formerly a senior consultant of James R. Knowles for his help in preparing many of the diagrams used for illustration.

Thanks are also given to Professor Ivor H. Seeley and the publishers for their support and constructive advice on the preparation and production of all stages of this book.

Last, but not least, to my wife, Joan, for her tolerance and support during the long evenings and weekends that I have taken to write this book.

1 Brief History of Construction Contracts and Case Law

1.1 Introduction

Modern contracts are used in a commercial environment which has encouraged the development of claims in construction contracts in recent years. Nevertheless, many of the conditions of contract used today are based on documents that were drawn up in the last century, and much of the construction law that is relied upon in the courts and in arbitration has been made as a result of cases that took place in the industrial revolution.

Civil engineering contracts evolved significantly in the nineteenth century, mainly as a result of the growth in transport, such as canals and railways. Most early contracts had the essential ingredients governing price, time for completion, damages and specification of the work to be done, but it was the construction of the canals and railways which eventually caused entrepreneurs to consider additional provisions such as health, safety and welfare and to make contractual provisions governing the requirements which were necessary to protect the workforce and the community. In his book *The Railway Navvies* (Penguin Books, 1981), Terry Coleman describes how the Chester and Holyhead Railway Company stipulated in contracts that the contractors should provide huts for the men where there was no room for them in the village along the line, and that the men should be paid on stated days in money, with no part paid in goods.

At the same time as the growth in civil engineering there was an increasing demand for buildings such as mills, factories and hostels for a working population which had flooded into the towns and cities. Building contracts had to take account of new pressures to complete on time, and new standards and specifications had to be drawn up to cope with new materials, such as cast iron, which were becoming available in commercial quantities. It is evident from reported cases throughout the nineteenth century that the roles of architect, or engineer or surveyor included that of an independent certifier when carrying out certain duties under construction contracts.

Gradually the contents of construction contracts became more sophisticated and included a host of new provisions; some brought about by Statute and others by the influence of the new professional institutions and trade associations that were being formed and which were to play an important role in a fast growing industry.

The method of tendering, in the early years of the industrial revolution, is best illustrated by Firbank, quoted by Coleman in *The Railway Navvies* (*supra*):

> 'Firbank himself used to tell a story of one Mr Wythes (probably George Wythes, who undertook, among other lines, that from Dorchester to Maiden Newton) who was thinking of submitting an offer for a contract. He first thought £18 000 would be reasonable, but then consulted his wife and agreed it should be £20 000. Thinking it over, he decided not to take any risk, so made it £40 000. They slept on it and the next morning his wife said she thought he had better make it £80 000. He did; it turned out to be the lowest tender notwithstanding, and he founded his fortune on it.'

Fortunes could be made quickly, but many contractors went broke due to underestimating the practical difficulties of constructing the work to strict standards in all weathers and a lack of awareness of the consequences of delay and other serious breaches of contract. It was soon realised that a major area of risk was inherent in the uncertainty of the quantity of work to be done and the variable ground conditions. Civil engineering contracts developed on the basis that all work would be remeasured at rates which were agreed at the outset; a reasonable solution bearing in mind the uncertainty of ground conditions which affected most of the work which was to be carried out. On the other hand, it was thought that building work was capable of quantification with reasonable accuracy (with the exception of changes ordered after the contract was agreed).

Therefore, building contracts were generally not subject to remeasurement and the contractor bore the risk of any mistakes which he may have made when measuring the work to be done from the drawings. The high cost of tendering for building work caused tendering contractors to engage a 'surveyor' who was responsible for measuring all of the work from the drawings and whose fees would be shared by all tenderers. Very soon this practice was overtaken by the employer (or his architect) engaging the surveyor to measure the work and for the 'quantities' to be provided for each tendering contractor for pricing the work. The surveyor's fees for measuring the work was usually required to be shown at the foot of the priced bill of quantities to be submitted with the tender and the successful contractor would then pay the surveyor out of the proceeds of interim certificates. This meant that each tendering contractor started by pricing the work based on the same bills of quantities, thereby reducing the cost of tendering and reducing the risk of error in quantifying the work to be done.

This practice, which survived for many years, caused problems if the building owner decided not to proceed with the work. Some building owners contended that they had no liability to pay the quantity surveyor's fees if the contract did not go ahead: *Moon* v. *Whitney Union* (1837), and *Waghorn* v. *Wimbledon Local Board* (1877); (*Hudson's Building and Engineering Contracts, tenth edition*, at pp 113 and 114). Even as late as the 1920s some standard forms of contract reflected this practice. The form of contract which was known by the short title as *The Model Form of Contract* (one of the RIBA publications referred to hereinafter), contained the following clause 14 prior to 1931:

> '(a) The fees for the Bills of Quantities and the Surveyor's expenses (if any) stated therein shall be paid by the Contractor to the Surveyor named therein out of and immediately after receiving the amount of the certificates in which they shall be included. The fees chargeable under clause 13 [Variations] shall be paid by the Contractor before the issue by the Architect of the certificate for final payment. (b) If the Contractor fails or neglects to pay as herein provided, then the Employer shall be at liberty, and is hereby authorised, to do so on the certificate of the Architect, and the amount so paid by the Employer shall be deducted from the amount otherwise due to the Contractor.'

Until 1963 the RIBA standard forms of contract contained optional provisions (clause 10) whereby the contractor could be responsible for paying the quantity surveyor's fees out of monies certified by the architect. However the quantity surveyor generally became engaged by the building owner, or his architect, who were responsible for paying the fees.

Whilst much of the case law which was relevant to construction contracts was shaped in the nineteenth century, there continued to be cases of note during the twentieth century. In parallel, non-standard and standard forms of contract evolved. The first 'standard forms of contract' were probably developed by public corporations. Revisions to many forms of contract were often prompted by decisions in the courts and these revisions (or the interpretation and application of them) sometimes became the subject of later cases which were to have a continuing influence on the draftsmen of new contracts and on the understanding of the law which affects contracts in construction.

Standard forms of contract which came into general use in building contracts were developed by the Royal Institute of British Architects (RIBA). By the early twentieth century the use of the RIBA form of contract was widespread. This form of contract, which was to be the subject of several editions and revisions, was to become the basis of most building contracts and was the forerunner of the Joint Contracts Tribunal (JCT) forms of contract of 1963 and 1980. In civil engineering, the first edition of the Institution of Civil Engineers (ICE) conditions of contract was launched in 1945. The fifth edition is currently in general use and the sixth edition (1991) is due to overtake its predecessor. One of the features of these standard forms of contract is that they are approved and accepted by the professional institutions and the contractors' associations. Several other standard forms of contract developed independently, such as GC/Works/1 for use by government departments and forms published by other professional bodies.

Internationally, particularly where there was British influence, standard forms of contract developed on the same lines as in the United Kingdom. Forms of contract which were (almost verbatim) the same as the RIBA/JCT forms of contract came into use in Cyprus, Jamaica, Gibraltar, Bahrain, Hong Kong and Singapore. Currently, in Cyprus, one of the first editions of the RIBA form of contract (probably used in the United Kingdom about the time of the First World War) is used alongside a variant of the 1963 edition of the JCT form of contract.

In Hong Kong a variant of the 1963 edition of the JCT form of contract is widely used and a draft based on the 1980 edition of the JCT form has been awaiting sanction since the early 1980s. Until recently, the form of contract used in Singapore was a variant of the 1963 edition of the JCT form. However, since 1980 the Singapore Institute of Architects has departed from following developments in the United Kingdom and has adopted an entirely new form of contract which bears no resemblance to any other standard form of contract used in the United Kingdom. In civil engineering a standard form of contract for use internationally was developed and agreed by the Federation Internationale des Ingenieurs-Conseils (FIDIC) using almost entirely the same format and conditions as the ICE conditions of contract. The second, third and fourth editions of FIDIC are currently being used in various parts of the world, often with extensive amendments beyond those contemplated by the Conditions of Particular Application in Part II of this form of contract.

1.2 Bills of quantities

Contractors who calculated their own quantities from drawings supplied by the building owner adopted methods of measurement according to their own style. The first quantity surveyors also prepared the bills of quantities in their own style and adopting their own particular methods of measurement. In the beginning this was probably confusing as the tendering contractors must have placed their own interpretation of the method of measurement. No doubt the quantity surveyors gradually developed methods which were fairly consistent and contractors became familiar with each individual quantity surveyor's method of measurement. The courts dealt with many cases involving liability for inaccurate bills of quantities and the decisions appear to be inconsistent. The apparent inconsistency was due in part to the distinguishing features of the various contracts and representations which were made regarding the quantities. However, it was held in *Bolt* v. *Thomas* (1859), (*Hudson's Building and Engineering Contracts, tenth edition,* at page 196) that where it was stipulated that the builder should pay the architect for the calculation of the quantities, and he had done so, then the builder was entitled to compensation from the architect if the bill was not reasonably accurate.

As late as the 1920s the Model Form of Contract (RIBA) did not incorporate a standard method of measurement, nor did it expressly state that the bills of quantities was a contract document. Nevertheless it was implied that the bills of quantities had contractual status and the contract contained provisions in clause 12a as follows:

> 'Should any error appear in the Bills of Quantities other than in the Contractor's prices and calculations, it shall be rectified, and such rectification shall constitute a variation of the Contract, and shall be dealt with as hereinafter provided.'

The provisions in the above contract have survived to the present day and almost identical wording appears in the 1963 and 1980 editions of the JCT form of contract. Similar provisions also appear in the fifth and sixth editions of the ICE conditions of contract in clause 55(2).

In the absence of a standard method of measurement, errors in composite descriptions and alleged omissions of items, as opposed to errors in measurement, became a constant source of argument. The first steps to rectify these difficulties probably took place in 1909, when the Quantity Surveyors' Association appointed a committee to prepare and publish pamphlets recommending the method of measurement for three trades. The first edition of the Standard Method of Measurement (SMM) was published in 1922 with the agreement of representatives of the Surveyors' Institution, the Quantity Surveyors' Association, the National Federation of Building Trades Employers and the Institute of Builders. The situation which existed prior to the publication of the the first edition is perhaps best described in the opening paragraph of the preface to this historic document:

> 'For many years the Surveyors' Institution and the Quantity Surveyors' Association (which bodies are now amalgamated) were accepted as the recognised authorities for deciding disputed points in connection with the measurement of building works. The frequency of the demands upon their services for this purpose directed attention to the diversity of practice, varying with local custom, and even with the idiosyncrasies of individual surveyors, which obtained. This lack of uniformity afforded a just ground of complaint on the part of contractors that the estimator was frequently left in doubt as to the true meaning of items in the bills of quantities which he was called upon to price, a circumstance which militated against scientific and accurate tendering.'

As might be expected, it took several years for the quantity surveying profession to become aware of the SMM and to use it in practice. Several years after the publication of the first SMM, in *House and Cottage*

Construction, Volume IV, Chapter II (Caxton Publishing Company Limited), Horace W. Langdon Esq., F.S.I, a practising Chartered Quantity Surveyor, made no reference to a standard method of measurement and he described how the quantity surveyor ought to explain the method of measurement used to prepare the bills of quantities.

The second edition of the SMM was published in 1927, and in 1931 the RIBA published its revised form of contract which (in clause 11) incorporated the SMM, where quantities formed part of the contract. The first test as to the valid incorporation of the SMM into the contract and the application and interpretation of the principles laid down in the standard method of measurement took place in 1938: *Bryant and Sons Ltd* v. *Birmingham Saturday Hospital Fund* [1938] 1 All ER 503. It was held that clause 11 of the contract, and the SMM, had been incorporated into the contract and that the contractor was entitled to extra payment for excavation in rock which ought to be measured separately pursuant to the principles laid down in the SMM.

It is evident that the decision in the Bryant case turned on the special wording in the standard form in clause 11, to the effect that the bills *unless otherwise stated* should be deemed to have been prepared in accordance with the current standard method of measurement. Almost identical provisions appear in clause 12(1) of the 1963 edition and in clause 2.2 of the 1980 edition of the JCT forms of contract and are the basis of many claims which persist in the construction industry today. The development of more sophisticated standard methods of measurement, whilst desirable in many respects, has done little to eliminate this type of claim. The provisions of SMM6 require the quantity surveyor to provide more detailed information than that required by the SMM (where necessary) (A1) and for the employer to provide information on groundwater (D3.1) or to state what information is assumed (D3.2).

Civil engineering quantities developed along similar lines to building quantities and standard methods of measurement became incorporated into contracts for civil engineering work. Clause 57 of the fifth and sixth editions of the ICE conditions of contract contains similar provisions regarding the status and application of the Civil Engineering Standard Method of Measurement (CESMM) referred to therein. Any work carried out by the contractor which is not measured separately in accordance with the CESMM may (unless there is a statement to the contrary) be subject to a claim for additional payment: *A.E.Farr Ltd* v. *Ministry of Transport* (1965) 5 BLR 94.

1.3 Variations

Building and civil engineering contracts are of such a nature that it is almost impossible, especially where work has to be carried out in the ground, to design and construct a project so that the final product is identical in every way to the original design which formed the basis of the contractor's tender. Changes to the original design and/or details may come about for technical reasons or because the building owner desires a revision to the plans or details.

Where technical reasons are the cause of a variation (for example, unsuitable ground conditions) the employer, or his architect, or engineer, will have limited control over the scope of the change in the work to be done by the contractor. Where the employer desires a change to the plans or details (for example, for aesthetic, or practical, or financial reasons), the scope of the change is to a large extent within the control of the employer. Without a suitable provision in a contract which allows the works to be varied, such changes would not be permitted (under the terms of the contract) and in the event of unavoidable changes for technical reasons the contractor would no longer be obliged to complete the work. Changes could only be executed by the agreement of the contractor or by way of a separate contract.

The standard forms of contracts used in building and civil engineering forms of contract provide for variations which are necessary or desirable (the latter being the employer's prerogative, but it does not exclude variations initiated by the contractor). The JCT forms of contract expressly provide for the architect to sanction a variation made by the contractor without an instruction issued by the architect.

Sometimes arguments are raised concerning the limit beyond which it may be regarded that the changes were outside the scope of the variation clause. Such arguments, if successful, would enable the contractor to refuse to execute the revised works or to escape from the contract rates and recover on a *quantum meruit* basis (a reasonable valuation in all the circumstances). There are no finite guidelines to assist in this matter. Some early forms of contract expressly stated a percentage of the contract price as the yardstick for determining the extent of variations permitted under the terms of the contract. The international form of contract (FIDIC) provides for a limited revision to the contract price if the sum total of all changes and remeasurement (with some exceptions) exceeds 10 per cent (clause 52(3) of the third edition) or 15 per cent (clause 52.3 of the fourth edition). However, this cannot be construed as being a true valuation on a *quantum meruit* basis. In the absence of stated limits such as a percentage, it is necessary to decide whether or not the scope of the changes went beyond that which was reasonably contemplated by reference to the contract documents and the surrounding circumstances of the case.

In *Bush* v. *Whitehaven Port and Town Trustees* (1888) 52 JP 392, the contractor was to lay pipes and possession of the site was to be given to the contractor for the performance of the work. Owing to delay in giving possession of the site to the contractor, the work had to be done in the winter, whereas it was contemplated that the work would be done in the summer. It was held that the contractor was entitled to payment on a *quantum meruit* basis (a reasonable price for the work in all the circumstances).

Modern contracts contain variation provisions which are so wide that it may appear doubtful that any claim for payment on a *quantum meruit* basis would succeed. However, in *Wegan Construction Pty. Ltd.* v. *Wodonga Sewerage Authority* [1978] VR 67 (Supreme Court of Victoria), the contractor successfully claimed on a *quantum meruit* basis. This case is worthy of further consideration on the grounds that the contractual provisions for variation were very wide (being similar to the ICE fifth and sixth editions and FIDIC fourth edition) and is summarised in Chapter 5.

Another problem which has come before the courts over the years, is the vexed question about omissions when the employer intends to have the work done by others. It is an increasingly common practice, when progress is delayed by the contractor, for the employer (through his architect) to omit work. This is often work which ought to be done by nominated subcontractors under the architect's instructions and its omission appears to be aimed at holding the contractor liable for liquidated damages (due to the contractor's own delay) on the mistaken premise that such an omission is a valid variation.

Presumably the employer believes that if the work is omitted, the architect does not have to issue any (late) instructions to carry out the work, which would have the effect of defeating the employer's claim to liquidated damages. It is well established in law that the power to omit work, even where the contract provides that no variation should in any way vitiate or invalidate the contract, is limited to genuine omissions, that is, work not required at all. It does not extend to work taken out of the contract for it to be done by another contractor: *Carr* v. *J. A. Berriman Pty Ltd* (1953) 27 ALJR 237 (Aus).

1.4 Extensions of time and liquidated damages

An extension of time provision is inserted in a contract for the benefit of both the contractor and the employer. However, its insertion is primarily for the benefit of the employer. Without such a provision, once the employer had caused delay, the contractor would no longer be bound to complete the works by the contract completion date and the employer would no longer be able to rely on the liquidated damages provisions in the contract. These fundamental points are often not appreciated by employers or their agents who are responsible for making extensions of time, in spite of the fact that decisions in the courts spanning almost two centuries have consistently reflected this view. In *Holme* v. *Guppy* (1838) 3 M & W 387, the contractors were responsible for delay of one week and the employer was responsible for delay of four weeks. There was no extension of time clause. It was held that the employer could not deduct liquidated damages from monies due to the contractor.

Draftsmen of contracts for building and civil engineering work recognised that there were many possible causes of delay to projects which were to be constructed over a period of years, in all weathers, and which were almost certainly going to be subject to delay by events within the control of the employer. Delays which were due to neutral events (such as inclement weather) and events which were generally within the control of the contractor were of no concern to the employer, and if contracts were delayed by such matters, then the contractor would have to take the necessary measures to make up the delay or face the consequences by payment of liquidated damages.

The use of contracts with onerous provisions which held the contractor liable for damages for every type of delay was not commercially satisfactory, as it encouraged cautious contractors to increase their prices and the reckless ones probably went out of business. Neither of these options were in the interests of the employer nor were they in the interests of the industry as a whole. On the other hand, delays on the part of the employer would extinguish the employer's rights to liquidated damages and it was therefore essential that the contract should include suitable provisions to enable the employer, or his agent, to make an extension in the event of delay for any cause which was within the employer's control or for which the employer was responsible (such as obtaining statutory approvals).

The drafting of suitable provisions which would protect the employer in the event of delay caused by him, and which would permit extensions of time for neutral causes and causes of delay which were generally within the control or at the risk of the contractor, proved to be a major problem. Very general provision such as 'circumstances wholly beyond the control of the builder' proved to be of no effect in circumstances where delay had been caused by the employer. This was held in *Wells* v. *Army and Navy Co-operative Society Ltd* (1902) 86 LT 764, where the extension of time clause contained the words 'or other causes of delay beyond the contractor's control'.

In spite of the decision in the *Wells* case (which was reported in the fourth edition of *Hudson's Building Contracts* in 1914), draftsmen of building and civil engineering contracts continued to use general terms which were almost certainly bound to be ineffective where the employer caused delay. Over fifty years later in *Perini Pacific Ltd* v. *Greater Vancouver Sewerage and Drainage District* [1967] SCR 189, delays of ninety-nine days occurred which included forty-six days on the part of the employer. The extension of time clause in the contract contained the provisions to extend time for completion due to 'extras or delays occasioned by strikes, lockouts, *force majeure* or other cause beyond the control of the contractor'. It was held that the extension of time clause did not cover delays caused by the employer and no liquidated damages could be recovered.

The fourth, fifth and sixth editions of the ICE form of contract and the third edition of FIDIC contain the general terms 'other special circumstances of any kind whatsoever'. It is evident, in view of the decisions in the *Wells* and *Perini Pacific* cases, that these standard forms of contract, some of which are still in use today, do not cover delay by the employer (with the exception of certain specified 'other cause of delay referred to in these Conditions'). It is conceivable that several causes of delay by the employer could occur in a civil engineering contract, which delays are not expressly covered elsewhere in the contract and which would therefore deprive the employer of its rights to deduct liquidated damages.

For many years standard forms of building contract appear to have been drafted in recognition of the difficulties caused by the *Wells* decision. Since the early part of this century the RIBA forms of contract have listed several causes of delay within the control of the employer (and other causes of delay) for which an extension of time could be granted. However, unless such a list is comprehensive, any delay which is not included therein would not qualify for an extension. If the non-qualifying delay was the employer's responsibility, no extension could be granted and the employer's rights to deduct liquidated damages would be extinguished. This point was clearly emphasised in *Peak Construction (Liverpool) Ltd* v. *Mckinney Foundations Ltd* (1970) 1 BLR 111. In this case a subcontractor (Mckinney) was guilty of defective work in the piling for foundations as a result of which there was a suspension of work. The subcontractor submitted design proposals to remedy the defects. The employer (Liverpool Corporation, a local authority) took an unreasonably long time to approve the subcontractor's proposals and the contractor was unable to continue with the works until some fifty-eight weeks later. The employer deducted liquidated damages for the period of delay and the contractor sought to recover the damages from the subcontractor. The contract contained an extension of time clause which set out the causes of delay for which an extension of time could be made, but it did not cover the employer's delay in approving the subcontractor's proposals. It was held that since part of the delay was due to the employer's default, and since there was no applicable extension of time provision, the employer could not deduct liquidated damages and he was left to recover such damages as he could prove flowed from the subcontractor's breach.

More recently in the case of *Rapid Building Group Ltd* v. *Ealing Family Housing Association Ltd* (1984) 29 BLR 5, the contractor was prevented from having full possession of the site on the due date. The contract was the 1963 edition of the JCT standard form of contract. There was delay and the works were completed late. The architect extended time for completion and issued a certificate that the works ought reasonably have been completed by the extended date for completion. The employer deducted liquidated damages for the period after the extended date for completion until the date when the contractor completed the works. It was held that the 1963 edition of the JCT form of contract did not provide for extensions of time due to the employer's breach of contract in failing to give possession of the site in accordance with the terms of the contract and the employer could not deduct liquidated damages from monies due to the contractor. The 1980 edition of the JCT form of contract includes failure to give possession of the site as a cause of delay (a relevant event) for which an extension of time may be granted.

Recent drafting (such as the fourth edition of FIDIC, GC/Works/1 and the Singapore Institute of Architects forms of contract) includes a list of causes of delay for which an extension of time can be made and there is a 'catch-all' provision intended to cover 'any act or default of the employer'. It is unlikely that this type of catch-all provision will enable the employer to cause delay with impunity. Some delays may well be beyond the contemplation of such a clause and the contractor may have grounds to determine his employment.

Even if a contract contains an effective extension of time clause, the employer's rights to deduct liquidated damages may be extinguished if the power to extend time for completion is not exercised within the time contemplated by the contract terms. In *Miller* v. *London County Council* (1934) 151 LT 425, the contract contained the following terms:

> 'it shall be lawful for the engineer, if he thinks fit, to grant from time to time, and at any time or times, by writing under his hand such extension of time for completion of the work and that either prospectively or retrospectively, and to assign such other time or times for completion as to him may seem reasonable'.

The contractor completed the works on 25 July 1932 and, on 17 November 1932, the engineer extended time for completion to 7 February 1932 and certified that liquidated damages were payable for the period from 7 February to 25 July 1932. It was held that the extension of time clause empowered the engineer to look back (retrospectively) at the delay as soon as the cause of the delay had ceased to operate and to fix a new completion date 'within a reasonable time after the delay has come to an end' (Du Parcq, J. quoting from *Hudson on Building Contracts, sixth edition* at page 360). The power to grant an extension of time had been exercised too late and the employer could not rely on the liquidated damages provision in the contract.

In another case, *Amalgamated Building Contractors* v. *Waltham Holy Cross UDC* [1952] 2 All ER 452, the contract was an RIBA form of contract which contained the following provisions in clause 18:

> 'If in the opinion of the architect the works be delayed...(ix) by reason of labour and materials not being available as required... then in any such case the architect shall make a fair and reasonable extension of time for completion of the works...'.

In this case the contractor was delayed due to non-availability of labour and during the month prior to the contract completion date he made two applications for an extension of time which the architect formally acknowledged. The date for completion was 7 February 1949 and the contractor completed the works in August 1950. In December 1950 the architect made an extension of time to May 1949. The contractor argued that an extension of time cannot be made to a date which has passed and therefore the extension was given too late. It was held, distinguishing *Miller* v. *London County Council*, that the extension of time could be made retrospectively and the extension was valid.

The different decisions in the *Miller* and *Amalgamated Building Contractors* cases are due to several distinguishing matters which are relevant. In *Miller* the engineer's decision on extensions of time was final and the wording in the two contracts were not the same. Perhaps more importantly, the cause of delay in *Miller* was within the control of the employer, whereas in *Amalgamated Building Contractors*, the cause of delay was beyond the control of the employer. In the latter case the delay was continuous, over a period of several months, thereby making it difficult, if not impossible, to estimate the length of the delay until the works had been completed. A detailed explanation of the law as it applies to this subject is given in the judgement in *Fernbrook Trading Co. Ltd* v. *Taggart* [1979] 1 NZLR 556. (For an excellent summary of this case, refer to *A Building Contract Casebook* by Dr Vincent Powell-Smith and Michael Furmston at page 355).

Contractors seeking to argue that the contract does not provide for extensions of time (for delay by the employer), or that an extension of time was made too late, thereby being invalid, may not necessarily be in a better position than they might have been by accepting a reasonable extension of time, valid or otherwise. If the contractor's arguments are successful the contract completion date is no longer applicable, the contractor's obligation is to complete within a reasonable time (time is at large) and the employer cannot rely on the liquidated damages provision to deduct the sums stated in the contract. In these circumstances the contractor does not have all the time in the world to complete the works, nor does he escape liability for general damages which the employer may suffer as a result of delay within the control of the contractor. Nevertheless, contractors may find it attractive to escape from the contractual period and the potential liability for delay at the rate stated as liquidated damages in the contract on the basis that the burden of proof shifts from the contractor to the employer. In *Wells* v. *Army and Navy Co-operative Society* (*supra*), Wright, J, the trial judge said:

> 'The defaults were, in my opinion, sufficiently substantial to cast upon the defendants [the employer] the burden of showing that the defaults did not excuse the delay.'

and in *Peak Construction (Liverpool) Ltd* v. *Mckinney Foundations Ltd*, (*supra*) Salmon, L.J. said:

> 'If the failure to complete on time is due to fault of both the employer and the contractor, in my view, the clause does not bite. I cannot see how, in the ordinary course, the employer can insist on compliance with a condition if it is partly his own fault that it cannot be fulfilled:...I consider that unless the contract expresses a contrary intention, the employer, in the circumstances postulated, is left to his ordinary remedy; that is to say, to recover such damages as he can prove flow from the contractor's breach.'

It is often argued that the employer cannot recover more in general damages than he would have been able to recover by way of liquidated damages. It appears from *Rapid Building Group Ltd* v. *Ealing Family Housing Association Ltd* (*supra*), that if the employer has lost his rights to liquidated damages, his claim for general damages may not be limited by the amount specified in the contract for liquidated damages. This point was not decided in the *Rapid Building* case but it must be at least arguable that this may be the case in certain circumstances.

In *Temloc Ltd* v. *Erril Properties Ltd* (1987) 39 BLR 31, the sum specified for liquidated damages was '£nil' and the employer sought to recover unliquidated damages arising out of delay in completion by the contractor. The Court of Appeal decided that by inserting a £nil rate for liquidated damages (to be calculated pursuant to clause 24.2.1 of a 1980 edition of the JCT form of contract), the parties had agreed that there should be no damages for late completion. However, in this case the Court of Appeal took the view that an extension of time which had been made by the architect after the twelve-week period required by clause 25.3.3 of the contract did not invalidate the liquidated damages provision and general damages could not be recovered as an alternative. Accordingly, the matter of the employer's rights in the event of the liquidated damages provisions being inapplicable did not have to be considered.

Nevertheless, notwithstanding the *Temloc* case, it appears likely that in the event of the contractor successfully arguing that the liquidated damages provisions are no longer applicable, then he may run the risk of being liable for general damages in excess of the liquidated damages. On the other hand, an employer who caused the liquidated damages provision to be invalidated, for any reason, for the purposes of claiming a higher amount of general damages than he might have recovered under the contractual provisions would be unlikely to find favour in the courts (see further commentary on the *Temloc* case in Chapter 7). This practice would surely fall foul of the rule of law which prevents a party from taking advantage of his own wrong, *Alghussein Establishment* v. *Eton College* [1988] 1 WLR 587.

Another vexed question arises in contracts where the employer intends to have phased completion and where the form of contract (usually a standard form) does not deal properly with this issue. In *Bramall and Ogden* v. *Sheffield City Council* (1983) 29 BLR 73, the contract incorporated the 1963 JCT conditions with liquidated damages 'at the rate of £20 per week for each uncompleted dwelling'. Extensions of time were granted but the contractor contended that further extensions were due and he disputed the employer's rights to deduct liquidated damages. The arbitrator awarded £26 150 as liquidated damages. On appeal it was held that the contract did not provide for sectional completion and the employer could not deduct liquidated damages.

It will be seen from the cases referred to that extensions of time and liquidated damages provisions in contracts merit careful drafting and that the interpretation placed on many provisions is open to dispute at almost every turn. The courts have generally taken a very strict view and the *contra proferentem* rule has usually been applied, (that is, the clause is usually construed against the interests of the party putting forward the clause and seeking to rely on it), *Peak Construction (Liverpool) Ltd* v. *Mckinney Foundations Ltd* (*supra*), and *Bramall and Ogden* v. *Sheffield City Council* (*supra*). The *contra proferentem* rule will not necessarily apply to contracts using standard forms such as the ICE or JCT forms of contract, *Tersons Ltd* v. *Stevenage Development Corporation* (1963) 5 BLR 54. The rule may be applied to particular amendments to a standard form imposed by the employer.

Extensions of time have perhaps been at the forefront of many disputes, most of which could have been avoided by care and attention to the matters which have been considered by the courts over many years. Later chapters will deal with some of these matters in greater detail.

1.5 Claims for additional payment: damages

Whenever there is delay, disruption or a change in circumstances or in the scope of the work, there is bound to be an effect on expenditure or income, either for the contractor or for the employer, or both. Subcontractors may also be affected. In some cases the risk is borne by the contractor (or subcontractor) and in others it may be borne by the employer. Where there is a breach of contract, or where there is a contractual provision to claim loss or damage, one party may have a claim against the other.

Claims relating to ground conditions are a regular feature in many building and civil engineering contracts. Numerous disputes have arisen as to the responsibility for information provided by the employer and upon whom the risk lies for unforeseen ground conditions. In *Boyd & Forrest* v. *Glasgow S W Railway Company* [1914] SC 472, the tendering contractors had only two weeks in which to tender for the work. The employer provided access to some information obtained by way of site investigations. The contractors claimed compensation for the losses caused by ground conditions which were not in accordance with the soil investigation information provided by the employer. It was held that the contractors were entitled to rely on the information provided by the employer and that the employer could not be protected against his own misrepresentation.

If employers were able to place the risk entirely on the contractor, the likelihood would be that tender prices would be much higher than if the risk was on the employer. The ICE and FIDIC forms of contract, being forms generally applicable to civil engineering contracts where a considerable amount of work is carried out in the ground, have provisions which recognise the problems associated with the uncertainty of ground conditions. Clauses 11 and 12 of these forms of contract have, in various editions over the years, provisions such as (quoting from the fifth edition of the ICE form of contract):

> '11 (1) The Contractor shall be deemed to have inspected and examined the Site and its surroundings and to have satisfied himself before submitting his tender as to the nature of the ground and sub-soil (so far as it is reasonably practicable and having taken into account any information in connection therewith which may have been provided by or on behalf of the Employer) the form and nature of the Site, the extent and nature of the work ...and in general to have obtained for himself all necessary information (subject as above-mentioned) as to the risks contingencies and all other circumstances influencing or affecting his tender.'
>
> '12 (1) If during the execution of the Works the Contractor shall encounter physical conditions (other than weather conditions or conditions due to weather conditions) or artificial obstructions which conditions or obstructions he considers could not reasonably have been foreseen by an experienced contractor and the Contractor is of the opinion that additional cost will be incurred which would not have been incurred if the physical conditions or artificial obstructions had not been encountered he shall if he intends to make any claim for additional payment give notice to the Engineer...'.

[The contract goes on to provide for an extension of time and additional payment.]

The above provisions appear to be a fair and reasonable attempt to ensure that contractors do not take the risk of *unforeseen* ground conditions and that employers are not exposed to unlimited claims. Notwithstanding these provisions, differences of opinion, ambiguity and deliberate tendering tactics have continued to provide an abundance of disputes and the results have often been against the interests of employers. Attempts have been made by the employer to escape responsibility for information on ground conditions provided by him.

In *Morrison-Knudsen International Co Inc and Another* v. *Commonwealth of Australia* (1980) 13 BLR 114, the employer disclaimed responsibility for the site investigation which he provided. It was held that the contractor was entitled to rely on the information provided and that the provisions in the contract were not an effective disclaimer. There may be a duty of care on the part of the employer in providing such information and the contractor may have a claim for misrepresentation, *Howard Marine & Dredging* v. *Ogden* (1978) 9 BLR 34.

Building contracts, by their nature, tend to be less vulnerable to claims involving ground conditions, but as can be seen from *Bryant & Sons Ltd* v. *Birmingham Saturday Hospital Fund* (*supra*), claims do arise from time to time.

The forms of contract in civil engineering recognised the concept of claims at an early stage and express provisions for additional payment in certain circumstances was a feature in these forms. The ICE conditions of contract use the term 'claim' whereas the RIBA and JCT forms of contract generally do not. Early RIBA forms of contract did not expressly provide for any additional payment over and above the contract rates except where it was appropriate under the variation clause. In the late 1920s and early 1930s the RIBA Model Form of Contract in general use contained no express provisions for 'delay and disruption claims' unless they could be dealt with as variations. Nevertheless it appears that architects and quantity surveyors of the time were of the opinion that there was power to make payment to the contractor without a variation being ordered. Horace W. Langdon Esq., F.S.I., wrote in *House and Cottage Construction (supra)*:

> 'EXTRAORDINARY CIRCUMSTANCES
> At times during the progress of work, certain happenings may take place which involve the contractor in a much greater expense than he had anticipated, such as, for instance, not being given a clear site, as may have been first promised. Under such circumstances, it is obvious that the cost per unit of the particular work affected must be greater than would have been the case had he had a clear run. Such a matter cannot be dealt with by the quantity surveyor, whose business it is to ascertain actual measurements of work executed and to value same as previously described. Extraordinary happenings of the kind mentioned would be dealt with by the architect. If the contractor disagrees with the architect's ruling, he may have recourse to the clause appertaining to arbitration.'

The RIBA form of contract referred to by Langdon did not contain provision for the extra payment which appears to be contemplated, nor did it provide for an extension of time for the breach of contract which was used as the example to explain 'extraordinary circumstances'. Misunderstanding of forms of contract and the application of the law persists today and is one of the reasons for disputes and actions for negligence.

The 1939 RIBA form of contract did not contain any provisions intended to deal with failure to give possession of the site or other acts of prevention by the employer, but it did contain new express provisions for additional payment in clause 1:

> 'If compliance with Architect's Instructions involves the Contractor in loss or expense beyond that provided for in or reasonably contemplated by this contract, then, unless such instructions were issued by reason of some breach of this contract by the Contractor, the amount of such loss or expense shall be ascertained by the Architect and shall be added to the Contract Sum.'

Provisions of the type quoted above are to be found in later editions of the RIBA and JCT forms of contract. Bearing in mind the wide rules for valuing variations where there are changes in circumstances, this type of provision appears to be intended to deal with the consequential effects of architect's instructions on other work (which work may not in fact have been varied by an instruction). This type of claim which involves delay and/or disruption to the regular progress of the works is troublesome for a variety of reasons that will be dealt with in later chapters.

One important ingredient of delay claims is often interest or finance charges. As a general rule this head of claim did not succeed unless it could be dealt with as special damages. The most important cases which deal with this matter came before the courts fairly recently and are discussed in later chapters. However, as modern disputes sometimes take years to settle, or to be decided, interest on the claim itself is often the largest single element of it. Where interest is awarded in favour of the contractor, a nominal amount over and above the bank rate is usually the measure of damages. The benefit to the employer however is often the return earned by 'turning the money over several times per annum' which, even in a moderately profitable business, may be up to ten times the amount of interest awarded. This level of damages is not contemplated, but it is perhaps difficult to reconcile this fact with the 'absolute rule of law and morality which prevents a party taking advantage of his own wrong whatever the terms of the contract: *Alghussein v. Eton College (supra)*.

An interesting feature of the 1939 edition of the RIBA form of contract was an optional clause (24(d)[A]) which provided for the retention fund to be deposited in a joint account in a bank named in the appendix to the contract. The interest which accrued was for the benefit of the employer, but as this was small compared with the return which could be gained by using the sum retained in a profitable business, the incentive for unscrupulous employers to seek to delay the release of the retention fund was reduced.

The more recent contracts issued by the JCT (JCT63 and JCT80) provide for the retention to be placed in a trust fund. This will provide a level of protection for contractors and nominated subcontractors in the event of the employer's liquidation and it will prevent employers using retention funds as working capital. At the outset of every contract, contractors should ask employers for details of the trust fund and ensure that all retentions are held in the said fund.

A number of recent cases have shown that contractors are being more cautious and are insisting on retentions being placed in a trust fund. If employers resist, the courts may issue an injunction to compel them to place the retention fund in a separate account: *Wates Construction (London) Ltd* v. *Franthom Property Ltd* (1991) 53 BLR 23.

1.6 Rolled-up claims

It is generally a requirement that the party making a claim should be able to illustrate that the damages claimed were caused by an event or circumstance which was a breach of contract or that it was a matter for which there was an express provision in the contract to make a payment therefor. It is not surprising that in complex building and civil engineering contracts, where many delays are occurring at the same time, it is difficult to allocate any particular element of damages to the appropriate event or circumstance which caused the damages claimed. In order to deal with this difficult problem, it was no doubt a common practice to formulate a general claim in which all of the damages which arose as a result of many interrelated causes were pursued as a 'rolled-up' claim.

This practice was challenged in *J. Crosby & Sons Ltd* v. *Portland Urban District Council* (1967) 5 BLR 121. In this case there had been some forty-six weeks' overall delay to completion due to various causes of delay of which thirty-one weeks had been held by the arbitrator as being attributable to causes of delay for which the contractor was entitled to compensation. The arbitrator proposed to award a lump sum to compensate for the delay of thirty-one weeks and the employer appealed claiming that the arbitrator should arrive at his award by determining the amounts due under each individual head of claim. The form of contract was the ICE fourth edition. It was held that, provided the arbitrator did not include an element of profit in the amount awarded, and that there was no duplication, then if the claim depends on 'an extremely complex interaction in the consequences of various denials, suspensions and variations, it may well be difficult or even impossible to make an accurate apportionment of the total extra cost between the several causative events', then the arbitrator was entitled to make a lump sum award for the delay and disruption.

This type of claim appeared in the case of *London Borough of Merton* v. *Stanley Hugh Leach Ltd* (1985) 32 BLR 51, where the form of contract was the 1963 edition of JCT. The judge was persuaded to allow a rolled-up claim on the basis of the findings in the *Crosby* case.

In a recent case, *Wharf Properties Ltd and Another* v. *Eric Cumine*

Associates, and Others (1988) 45 BLR 72, (1991) 52 BLR 1 PC, the employer (*Wharf*) pursued a rolled-up or global claim against his architect (*Cumine*) which relied on the same premise as both the *Crosby* and *Merton* cases. The Court of Appeal of Hong Kong did not accept the claim. On the face of it, there appears to be an anomaly which places doubt on the validity of this type of claim. However, in this case, there appears to have been a lack of evidence to link the damages claimed with the numerous alleged defaults of the architect. The *Wharf* case should not be regarded as the death knell for all claims of this kind. It should be noted that the judge in a subsequent case, *Mid-Glamorgan County Council* v. *J Devonald Williams & Partner*, 17 September 1991 (unreported), considered the previous cases involving rolled-up claims (including the *Wharf* case) and held that, provided the circumstances were appropriate, such a claim could succeed.

1.7 Notice

Most building and civil engineering contracts contain provisions which require the contractor to give notice of delay or of its intention to claim additional payment under the terms of the contract. It is usual for the contract to specify that notice should be given within a reasonable time, but other terms such as 'forthwith', or 'without delay' or within a specified period of the event or circumstance causing delay or giving rise to the claim may be used. The courts have had to consider the meanings of various terms and they have often been faced with the argument that the giving of notice was a *condition precedent* to the contractor's rights under the contract.

The ICE conditions of contract generally opt for a specified period within which notice should be given. Two cases involving the ICE conditions of contract are helpful in deciding if notice is a *condition precedent*.

In *Tersons Ltd* v. *Stevenage Development Corporation (supra)*, the engineer issued a variation instruction for the first contract on 24 July 1951. The contractor carried out the varied work and gave notice of his intention to claim on 3 December 1951. In the second contract the engineer issued an instruction on 24 August 1951 and the contractor gave notice of his intention to make a claim on 6 February 1952. Work on the second contract commenced on 12 March 1952. The contractor did not submit his claims on a monthly basis.

The Court of Appeal was asked to decide whether the contractor's notices complied with the provisions of sub-clauses 52(2) and 52(4) of the second edition of the ICE conditions of contract. Sub-clause 52(2) required the contractor to give notice of his intention to claim a varied rate 'as soon after the date of the Engineer's order as is practicable, and in the case of additional work before the commencement of the work or as soon thereafter as is practicable.'

Sub-clause 52(4) provided for claims to be made monthly and 'no claim for payment for any such work will be considered which has not been included in such particulars. Provided always that the Engineer shall be entitled to authorise payment to be made for any work notwithstanding the Contractor's failure to comply with this condition if the Contractor has at the earliest practical opportunity notified the Engineer that he intends to make a claim for such work.' It was held that clause 52(2) only required a notice in general terms that a claim was being made and that clause 52(4) only related to payment in monthly certificates. The proviso in clause 52(4) which empowered the engineer to authorise payment, and the provisions of clauses 60, 61 and 62, which contemplated that the contractor's rights remained open until the final maintenance certificate had been issued were sufficient to show that the contractor had complied with the contractual provisions.

In *Crosby* v. *Portland U D C (supra)*, the works were suspended by order of the engineer and the contractor did not give notice within period specified in sub-clause 40(1) of the fourth edition of the ICE conditions of contract which contained the proviso 'Provided that the Contractor shall not be entitled to recover any extra cost unless he gives written notice of his intention to claim to the Engineer within twenty-eight days of the Engineer's order.' It was held that since the contractor had not given notice within the specified period the claim failed.

The distinction between the *Portland* and the *Crosby* cases is best explained in *Bremer Handelsgesell-Schaft M. B. H.* v. *Vanden Avenne-Izegem P. V. B. A* [1978] 2 Lloyds LR 109, in which Lord Salmon said:

> 'In the event of shipment proving impossible during the contract period, the second sentence of cl. 21 requires the sellers to advise the buyers without delay and the reasons for it. It has been argued by buyers that this is a *condition precedent* to the seller's rights under that clause. I do not accept this argument. Had it been intended as a *condition precedent*, I should have expected the clause to state the precise time within which the notice was to be served, and to have made plain by express language that unless notice was served within that time, the sellers would lose their rights under the clause.'

In the *Portland* case neither of the ingredients stated by Lord Salmon were present, whilst in the *Crosby* case both ingredients were present. If notice is to be a *condition precedent*, it is important to take account of these essential requirements when drafting the relevant provisions. However, in some circumstances, this may be self defeating (for example, extensions of time provisions for causes of delay within the control of the employer).

The requirements to give notice in RIBA and pre-1980 JCT standard forms of contract were less stringent than the requirements in the ICE conditions. Notice of delay under the extension of time clause (clause 23 in the 1963 edition of JCT) is required to be given by the contractor 'forthwith'. The case of *Merton* v. *Leach* (*supra*) dealt with a host of issues, one of which involved extensions of time if the contractor fails to give written notice upon it becoming reasonably apparent that the progress of the works is delayed. It was held that, if the architect was of the opinion that the progress of the works is likely to be delayed beyond the completion date by one of the specified causes of delay for which there was power to extend time for completion of the works, the architect owes a duty to both the employer and the contractor to estimate the delay and make an appropriate extension of time. The giving of notice of delay by the contractor was not a *condition precedent* to an extension of time. However, failure on the part of the contractor to give notice in accordance with the contract was a breach of contract and that breach may be taken into account when considering what extension should be made.

1.8 Interference by the employer

Most building and civil engineering contracts provide for the architect or engineer to be responsible for granting extensions of time and certifying payment of sums due under the contract. In carrying out these duties the architect or engineer is required to act fairly and impartially and the employer is not permitted to influence or obstruct them in the performance of their duties. Several early cases show that the courts have taken a consistent view in cases where the employer has sought to influence the person appointed by him to certify or value in accordance with the contractual provisions, even if there was no fraud on the employer's part,*Hudson's Building and Engineering Contracts, tenth edition* at pp 460 to 463. In the case of *Morrison-Knudsen* v. *B. C. Hydro & Power* (1975) 85 DLR 3d 186, all of the contractor's requests for an extension of time were rejected and no extensions of time which were due to the contractor were granted. The contractor accelerated the progress of the work and the project was completed shortly after the contractual date for completion. It was subsequently discovered that the employer was instrumental in securing an agreement with a government representative that no extensions should be granted. The Court of Appeal of British Columbia held that the contractor was entitled to recover the acceleration costs which he had incurred as a result of the breach of contract. Further, the contractor would have been entitled to rescind the contract and sue for payment in *quantum meruit* if he had been aware of the breach.

In a recent Scottish case, the contractor claimed to be entitled to interest on a sum which the contractor claimed to be due but which had not been certified by the engineer. The contract was the ICE fifth edition which provided for interest to be paid in the event of failure to certify (clause 60(6)).

The Judge held that the clause did not allow for interest if the engineer certified sums which were less than the sums which the engineer ultimately certified as being due. If the engineer had certified what in his opinion was due at the time, it could not be construed as a failure to certify.

However, it was discovered that the employer had instructed the engineer that under no circumstances should he certify more than a specified sum without the employer's permission. The engineer appeared to ignore the employer's instructions and prepared a draft letter to the contractor indicating that a sum exceeding the employer's ceiling was due. The employer sacked the engineer. The Judge held that the employer's interference was sufficient to deny effect to the engineer's certificates in which case there must have been a failure on the part of the engineer to certify within the meaning of clause 60(6) of the contract. In these circumstances the contractor was entitled to interest: *Nash Dredging Ltd* v. *Kestrell Marine Ltd* (1986) SLT 62. [This decision, on the general matter of interest payable in accordance with the provisions of clause 60(6) of the ICE conditions, should not be regarded as being applicable in England. See *Morgan Grenfell* v. *Sunderland Borough Council and Seven Seas Dredging Ltd (infra)* Chapter 5.]

1.9 Claims against consultants

It has long been held that if a consultant acts negligently in the performance of his duties, and the employer suffers loss as a result, then the employer would have a claim for damages against the consultant. This was held to be the case in *Sutcliffe* v. *Thackrah and Others* (1974) 4 BLR 16. It appeared from the judgement in this case that the contractor may have a claim for damages against the consultant.

Several cases involving claims by contractors against consultants have been reported and the industry seemed to have a clear picture of the law in this regard when the contractor in *Michael Salliss & Co Ltd* v. *E C A Calil and William F Newman & Associates* [1989] 13 ConLR 68, successfully claimed damages arising out of the architect's failure to exercise properly the duty of care owed to the contractor. The law, as it appeared after the *Michael Salliss* case, was turned upside down in *Pacific Associates Inc and Another* v. *Baxter and Another* (1988) 44 BLR 33. In this case the Court of Appeal rejected the contractor's claim for damages arising from the engineer's negligence. The contractor had settled with the employer and sought to claim against the engineer on the grounds that:

> 'By their continual failure to certify and by their final rejection of the claims the engineers acted negligently and alternatively were in breach of their duty to act fairly and impartially in administering the contract.'

As it now stands, contractors are unlikely to succeed in claims for damages against consultants if the claim is one which the contractor can make against the employer. The situation may be different if there is no arbitration clause in the contract.

1.10 The future

The law relating to construction contracts has evolved rapidly in recent years and it looks set to continue at a similar pace in the future. Recent cases have put new interpretations on some aspects of the law but many grey areas still exist. The wide range of new or revised forms of contract will bring with them new problems that will need resolution. An increasing awareness of contract law and its application in modern contracts will be in evidence and new contractual provisions will be drafted to deal with the decisions of the courts. A considerable effort needs to be made in the direction of contracts administration, monitoring progress, claims formulation and presentation, and this is likely to be evidenced by the ever increasing number of seminars and training courses on the subject.

Resolution of disputes has become an increasingly costly exercise where the costs of arbitration are often no less than the costs of litigation. Procedures, extensive pleadings, tactics and joining of several parties have been the cause of escalating costs of managing an arbitration. The use of Alternative Dispute Resolution (ADR) is bound to find favour with all sides of the industry if there is a willingness to find better and cheaper means of settling disputes.

The Single European Market and the changes which it will bring to the construction industry in the United Kingdom and throughout the Community will widen the scope for professionals, developers, contractors and subcontractors. EC legislation will affect certain aspects of building procurement. A broader understanding of different legal systems and methods of contracting will be necessary to capitalise on the opportunities that this new market can offer. Whatever the future holds, many aspects of law and contracts that have been considered herein will continue to form the backbone of the system used in the United Kingdom.

2 Choice of Contracts

2.1 The first steps

There are three main categories of client who require the construction of, or alterations, or extensions to, a building or civil engineering project. The first category consists of clients who embark upon a building or civil engineering venture only once or perhaps a few times. The second category consists of clients who regularly have the need to refurbish, alter or expand existing premises or develop new projects in the course of their business. The third category comprises a variety of speculative developers who construct projects for sale or lease.

Clients who embark upon any construction venture for the first time are often faced with a number of alternative routes but usually the first stop will be at the office of a qualified architect or engineer. For the majority of projects this approach may be sufficient. Most professional firms of architects and engineers are well versed in the use of standard forms of contract and, unless the client has unusual requirements, a standard form of contract will be available to suit most purposes. They are, however, not without their pitfalls and some architects and engineers fail to provide the necessary advice which may make the difference between ultimate client satisfaction and a potential claim for professional negligence.

Whether it is an architect, engineer, quantity surveyor, solicitor or a lawyer specialising in construction contracts, the best advice is usually given by someone who has had 'hands on' experience in administering or managing contracts and is well versed in contract law, including all of the recent developments in case law which affects the interpretation and application of standard forms of contract. An unamended standard form of contract may be more appropriate than a masterful piece of legal drafting which fails to take account of practical reality and commercial practice. In most cases a good contract will comprise the appropriate standard form suitably amended to rectify its deficiencies and incorporating reasonable client's requirements.

Clients who are familiar with the pitfalls of contracting often have their own amendments for use with a standard form or they may have a tailor-made form of contract to suit their own requirements. This is a step in the right direction but recent cases in the courts have shown that many amendments to tried and tested standard forms of contract, and some provisions in hybrid forms of contract, fail to contain the standard of clarity necessary to ensure that the draftsman's intentions are understood. The application of the '*contra-proferentem* rule' and other well established principles in English law may assist contractors when the terms of the contract are decided in the courts.

The criticism of contractual provisions introduced by major corporations and public clients suggests that some of them should approach the problems of contracting with equal caution to first time venturers. The vast sums of money which may be at stake merit special attention to the contract conditions and one of the first steps which ought to be taken by any client embarking on a major project should be to obtain expert professional advice from someone who is not a member of of its own organisation. If this is done, the incidence of provisions which may appear to be in the client's interests, but which are likely to have the opposite result, may be reduced.

Some clients may be advised to proceed on the basis of an outline design brief and contractors may be invited to tender for the design and construction of the project. Independent advice is essential at all stages if this is to be adopted. If the client has confidence in a particular contractor, it may choose to go directly to the contractor to negotiate for the design and construction of the project. Only in exceptional circumstances should a client contract for work in this manner without the guidance of an independent professional throughout the contract.

2.2 Clients' objectives

The principal objectives of any client will be to have the project completed on time, within budget and to an appropriate standard of design, workmanship and materials. The priority or emphasis placed on these objectives will depend on a number of factors. Cost or time may determine the scope for design and specification for the work.

In view of the commercial pressures to minimise finance costs and to obtain revenue at the earliest possible date, priority may have to be given not only to a method of construction which is conducive to speed of erection, but to 'lead-in' times, phasing of design and construction, phased completion of the project, design by contractor and subcontractors, installation of client's equipment and many other factors depending on the complexity of the project. Major subcontractors or packages of work may have to be settled in advance of selection of the principal (or main) contractor. If a client has a generous budget, he may insist on the best quality and design whilst cost and time are secondary.

Whatever the client's objectives it is important to set out a master programme, showing the various anticipated design and construction phases, at an early stage. This may have a bearing on the type of contracting methods to be used and should not be overlooked. The most common causes of construction delay claims stem from insufficient time allowed for design and commencing on site before sufficient design and detailing has been completed.

2.3 Contracting methods

The most common method of contracting is where a contractor undertakes to complete the project for a lump sum according to the design prepared by an architect or engineer at the outset. This 'traditional' method of contracting envisages the design being complete subject only to explanatory details and limited provisional items. Any change to the original design will be dealt with by way of a variation. The size and complexity of the project may determine whether or not bills of quantities are to be used. In building contracts the bills of quantities are not generally subject to remeasurement (except for correction of errors in the quantities). In civil engineering it is generally accepted that the design may be dependent on factors outside the control of the employer (ground conditions) and the contract is subject to remeasurement.

This method of contracting, by its nature, contemplates substantial completion of the design by the designer at tender stage. That is not to say that every detail has been drawn. It envisages issuance of details which do not change the original design, but merely explain more fully what is shown on the contract drawings. In the normal course of events, providing the designer had considered the details necessary to make the overall design fit together, explanatory drawings should not constitute a variation to the original design.

It is often the case that some critical aspects of design cannot be properly represented on a drawing before the designer has drawn the details. This is fundamental drawing practice. Due to pressure to get tender documents together at the earliest possible stage, too many contracts get off to a bad start due to insufficient attention to detail before invitations to tender. In short, this type of contract envisages a design phase which is almost complete before the construction stage commences, and the only design to be done after commencement of construction is of an explanatory nature and variations to the original design for which there is machinery to adjust the contract sum and/or the contract period. (See Figure 2.1.)

Support for the view that a lump sum contract should be designed in all its essential elements at tender stage is found in *The Banwell Report* (*The Placing and Management of Contracts for Building and Civil Engineering Work – HMSO 1964*). The JCT standard forms of building contract used for this method of contracting clearly contemplate the design being substantially, if not wholly, complete at tender stage. The recitals of the JCT forms expressly state that the employer 'has caused Drawings and Bills of Quantities showing and describing the work to be done to be prepared by ...'. Clause 1.3 of JCT80 defines these Drawings as The Contract Drawings, and clause 2.1 requires the contractor to 'carry out and complete the Works shown on the Contract Drawings ...'.

It has long been an accepted practice, and provided for in most forms of contract, that some work may not be fully designed at tender stage. This is usually dealt with by provisional sums or provisional quantities. In recent years the proportion of work covered by provisional items has increased beyond that for which this type of contract was intended. In some cases as much as forty per cent of the contract sum has been made up of provisional items, leaving the contractor unsure as to the scope of the work and the employer without a realistic budget for the project.

Other forms of abuse include the use of provisional sums under the guise of PC (Prime Cost) Sums. Very often the prime cost sum is no more than a provisional sum, whereas on the strict interpretation of the contract, a prime cost sum should be a reasonable estimate based on a design which was in existence at tender stage. This will be dealt with in more detail in later chapters.

Some practitioners are bent on using a form of contract intended for use in the above circumstances (such as JCT80), when it was known at the outset that the design stage would extend well into the construction phase. This practice may work if the designer co-ordinates the design into a master programme which is synchronised with the contractor's construction programme. However, there are many risks, such as under-estimation of 'lead-in' times for procurement, limitation on the flexibility in the contractor's programme (in the event that the contractor needs to change sequence for his own convenience) and an unacceptable incidence of variations caused by lack of foresight. All of these factors may lead to late completion and claims for compensation of one kind or another.

Another disadvantage of traditional contracting is that it does not usually permit the contractor to have an input at design stage. Many contractors are able to contribute to the design so that savings in cost and time can be made for the benefit of the employer. Sometimes contractors offer alternative designs, but very often this is so late in the day that it places more pressure on the design team to take account of the contractor's proposals in the overall design. Variants on the traditional forms of contract include an element of design by the contractor such as JCT80 used with the 'Contractor's Designed Portion Supplement (CDPS) 1981 (revised 1988)'.

It is becoming increasingly popular for employers to move in the direction of design and build or turnkey contracts. A degree of competition may be introduced by a comprehensive design brief and a schedule of the client's requirements. It is important to ensure that firms bidding for work of this

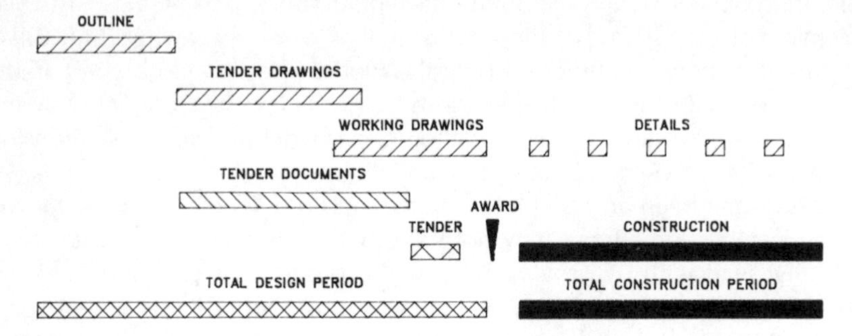

Figure 2.1 Traditional contracting

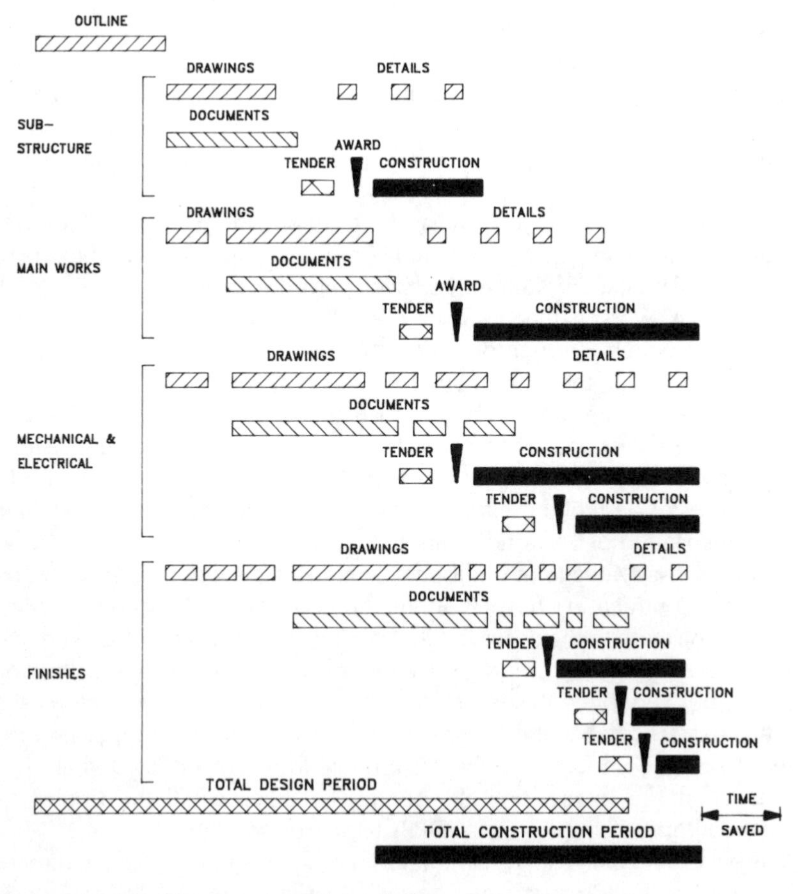

Figure 2.2 Phased design and construction

nature have a sound track record which can be verified and that a detailed inspection of previous projects is undertaken by the client's professional advisers. Care should be taken to investigate previous performance. Have the projects been completed on time and within budget? What are the maintenance costs? In addition to written testimonials from previous clients, it may be advisable to obtain permission to discuss the bidding contractors' performance and the quality of the buildings with clients and consultants for previous projects.

It is important to select a contractor in whom the client has complete faith and confidence. That is not to say that the client should go ahead without professional advice throughout the project. This may take the form of a project manager and possibly a quantity surveyor. An architect or engineer may also be engaged to advise on technical matters. A good project manager can make the difference between the success or failure of this method of contracting. It is essential that the person selected to carry out this role is given the freedom to act fairly and impartially. Whilst the employer's interests must be given priority, it is very often counter-productive to adopt an adversarial position which creates distrust between all parties. Much more benefits can be obtained for the client if the project manager helps to preserve trust and confidence by showing authority, integrity and competence at all levels.

There are circumstances in which it is advantageous for the design stage of the project to overlap with a considerable period of the construction phase (see Figure 2.2). If this is carefully structured, it is possible to commence construction much earlier than in traditional methods of contracting. The total effect of this method of contracting may be to give rise to a higher overall expenditure on construction: however, if the client can get beneficial occupation earlier than it otherwise would have done by traditional contracting, there may be considerable savings or benefits such as earlier rental income and reduced finance charges.

There are several methods of contracting which are suitable where it is intended that the design stage and the construction stage overlap. Management contracting is one method which lends itself to this process. In its purest form it is based on the prime cost plus the fixed (or percentage) fee method of contracting which has been used for many years. The outline design of the project, together with a detailed brief, is prepared by the design team and bidding contractors are required to submit their proposals for the management and 'procurement of construction'. The criteria used as a basis for selection will include:

- Reimbursable costs of site management, supervision and general services (similar to 'Preliminaries' in traditional contracting);
- Lump sum or percentage to be added to the prime cost of the project;
- Management capability and resources;

- Ability to contribute to the design of the project; 'buildability';
- Programme and methods of construction;
- Methods of ensuring quality control;
- Systems for cost control;
- Industrial relations;
- Proposed packaging of work to be done by subcontractors;
- Buying power and negotiation skills;
- Previous track record.

The selected management contractor does not usually execute any work himself. His obligations are, in collaboration with the design team and the employer, to procure completion of the project on time and within budget, by subcontracting various parts of the work and by purchasing materials to be fixed by subcontractors. Balance will have to be made when considering the size and scope of work packages.

Large packages will not enable the employer to obtain the benefit of buying margins, but a lower management fee may be required. On the other hand, a large number of small work packages will usually reduce the prime cost, but the management fee and reimbursable costs may be higher to reflect the increased management, supervision and risk involved.

In this method of contracting, the management contractor enters into an agreement with the employer in the same way as the contractor in traditional contracting. The contracting structure is shown in Figure 2.3. It is often the case that the management contractor's liability for late completion is limited to any damages which it can recover from subcontractors. This can cause serious problems if the subcontractors are financially vulnerable. Subcontractors carrying out small work packages may be faced with damages for late completion which are out of proportion to the value of work undertaken by them.

In traditional contracting, the employer may recover all of the damages from the contractor without being concerned about which subcontractors were the culprits. In management contracting, the liabilities of several subcontractors responsible for overlapping delays can cause difficulties and may often lead to disputes and arbitration or litigation.

Some hybrid forms of management fee contracts place greater responsibility on the management contractor. It is possible to devise a scheme where the management contractor is also responsible for the execution of the work in the same way as the traditional contractor. The advantages are that the management contractor is involved in the design and selection of subcontractors, but once the subcontracts are awarded, the management contractor takes full responsibility as if the subcontractor was a normal domestic subcontractor in the traditional sense. The management contractor may also execute some of the work himself. The management fee is likely to be higher to reflect the greater risk in this form of contracting.

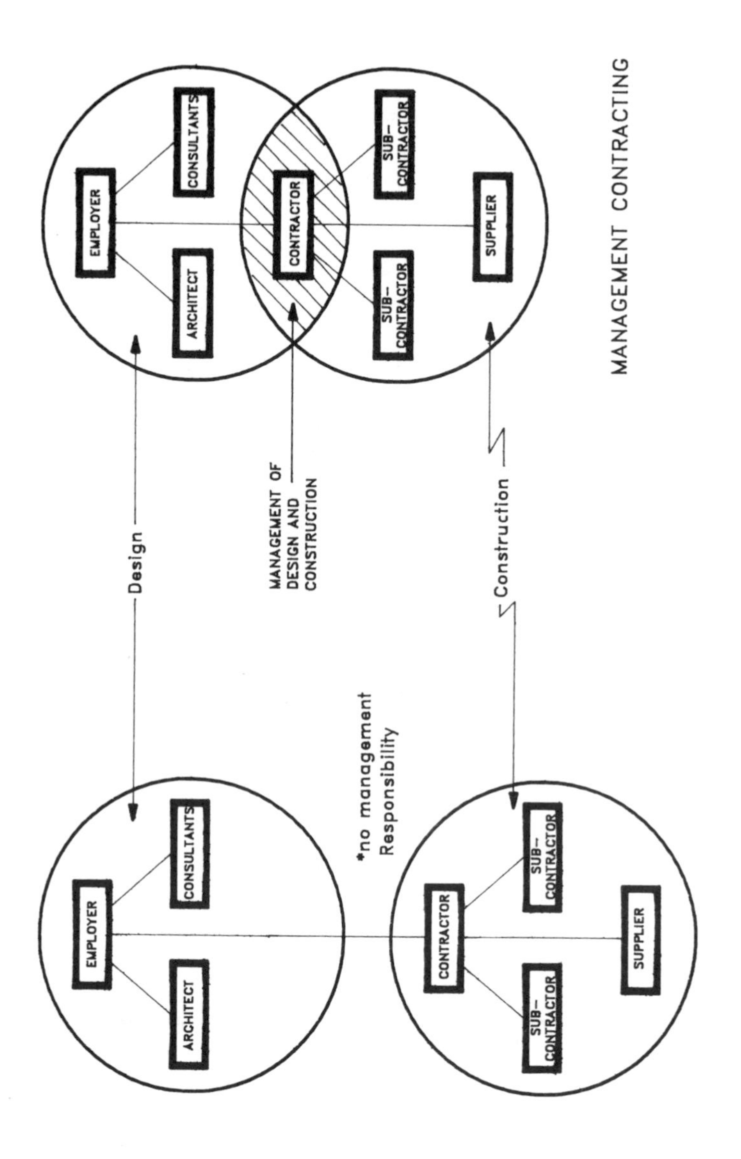

Figure 2.3 Management of design and construction

There are also many methods of project management or construction management which permit overlapping of design and construction. It is impossible to define these methods of contracting as there appears to be numerous variations on a theme. In very broad terms the project manager is responsible for co-ordinating and managing the design and construction of the project as part of the project team. The manager will enter into a contract with the client to manage the project, but he may not enter into subcontracts. Each work package is undertaken by direct contracts with the client and the work is carried out under the direction and supervision of the project manager (see Figure 2.4).

2.4 Standard forms of contract

Why use a standard form of contract? Firstly, it will have been prepared having regard to the nature of the work to be undertaken. Secondly, practitioners in the industry are more comfortable using a standard form of contract with which they are familiar and which is usually capable of interpretation by reference to readily available text books and case law. Thirdly, they are often drafted and agreed by recognised bodies representing

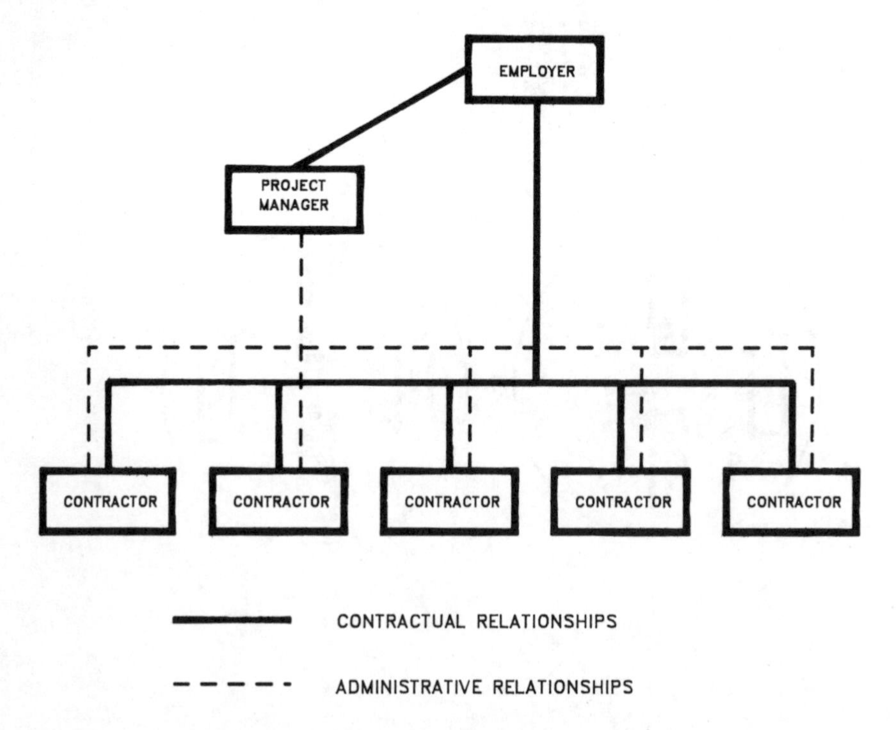

Figure 2.4 Project management structure

all sides of the industry which will be affected by them. This last point is to some extent a disadvantage in that a form of contract, 'by committee', is often a compromise containing some defective aspects of one form or another.

Standard forms of contract are available to suit contracts of almost any size and complexity and to suit most methods of contracting. Some practitioners select forms of contract with which they are familiar without having sufficient regard to their suitability or limitations. This practice is not to be recommended and should be regarded as 'short changing' the client. Any client embarking upon a construction project is entitled to expect sound advice from his professional advisers on all aspects of the contract, not least of which is the selection of the right form of contract for the purpose.

The methods of contracting discussed in this chapter will be a major consideration for many larger projects and for small or medium projects that require a considerable amount of preplanning. The type and size of contractors bidding for the job will also be important. For example, the use of a lengthy standard form, such as JCT80, may not be appropriate when the tendering contractors are little more than 'one man' firms having no understanding of the complicated provisions in the contract. The use of this form of contract in such circumstances will increase the price and/or lead to all sorts of problems in administration of the contract. At the other end of the scale, the use of one of the simpler forms of contract may not be appropriate for a project with a high building services content.

It is not possible to deal with all of the standard forms of contract in one chapter. However, some of the most common are considered very briefly.

2.5 The Joint Contracts Tribunal standard forms of contract

The most commonly known standard forms of contract are those issued by The Joint Contracts Tribunal (JCT). The first standard form issued by the JCT was in 1963 which superseded the RIBA forms of contract. It was published in four main variants; the private and local authorities versions, each with, or without, bills of quantities. Today there are a number of standard forms for a variety of needs.

The Minor Works Form, MW80

A simple form of contract embodying the essential ingredients of a building contract. Suitable for a project of limited value (not recommended for projects exceeding £70 000 at 1987 prices) where there are no bills of quantities. It is not suitable where nominated subcontractors are contemplated. The recommended limits on its use are contained in practice notes issued by the JCT. The practice notes are for guidance only and do not form part of the contract. As the title implies, the form is intended to be used for minor works which can be adequately defined in drawings and specification.

The Intermediate Form of Building Contract, IFC84

This form of contract was drafted to fill the gap between the minor works form and the standard form of building contract. It combines the simplicity of the minor works form of contract but many of the procedural provisions of JCT80 are incorporated. The same form can be used for private and local authorities' use, and it contains alternative provisions so that it can be used with a specification, or schedules of work or bills of quantities. Limitations as to its intended use are printed on the cover of the form of contract and further guidance is given in practice notes.

Supplementary conditions are provided if it is intended to have partial possession or sectional completion. Without these supplementary conditions, difficulties may arise when applying the liquidated damages provisions. Whilst this is a simplified form of contract when compared with JCT80, it is contained in more than thirty pages, making it almost as long as the predecessor of JCT80 (that is JCT63). With very little amendment, it is an extremely flexible form of contract which finds favour outside of its intended limitations.

The Standard Form of Building Contract, JCT80

Ignoring the fact that versions of its predecessor (JCT63) are still used in many parts of the world, this standard form of contract is perhaps the most widely used in building works today. Many aspects of JCT63 have been retained, including some which have received criticism in the courts over the years. Some of these will be discussed later. Provisions for dealing with nominated subcontractors have become unnecessarily complicated. Several amendments and practice notes have been issued. It is available in private and local authorities' editions with, or without, (bills of) quantities.

The JCT forms of contract referred to above are all intended to be used where the design has been substantially completed at tender stage. Other forms of contract issued by the JCT contemplate some of the design being a continuing process after tender stage (and after commencement of work). They include:

The Standard Form of Contract with Approximate Quantities

This form of contract may be suitable where the general contract philosophy of the JCT80 standard form of contract is to be retained but where the design is less complete than that required when using the standard form. It may be used if it is intended to bring forward the date of selection of a contractor with a view to earlier commencement on site. The quantities are subject to remeasurement. This contract is sometimes abused. It should not be a device to permit less accurate bills of quantities to be used.

The Fixed Fee Form of Contract

This contract may be suitable where the design has not progressed sufficiently to accurately define the *Works*. However, the scope of the work to be done has to be defined and sufficient information to describe the *items of work to be done* is necessary. An estimate of the prime cost of the work to be done and a fixed fee forms the basis of the estimated total cost to the employer. There is no provision to vary the scope of the work. The final cost to the employer is the actual prime cost ascertained from the contractor's accounts and invoices plus the fixed fee quoted by the contractor. There is provision for reimbursement of loss and expense caused by disturbance of the regular progress of the works.

The 1987 publication of this form of contract retains the format of the 1963 JCT standard form of contract. Some of its provisions, therefore, are subject to the same criticism as JCT63.

The Standard Form of Management Contract

The principle of ascertaining the cost to the employer, in this form of contract, is similar to the fixed fee form. The main differences between the fixed fee form and the management form are:

- The management contractor must co-operate with the design team as part of his contractual obligations;
- There is provision for a pre-construction stage and a construction stage;
- The management contractor does not carry out any work himself.
- In addition, there are optional contractual provisions dealing with instructions involving acceleration or revised sequence of work.

Control of cost and time is dependent upon the close co-operation between all members of the design team and the management contractor. The management contractor manages and supervises the construction of the work and the execution is done by several works contractors.

The Standard Form of Building Contract with Contractor's Design

This form of contract contemplates a reasonably detailed outline of the employer's requirements based upon which competitive tenders are invited, incorporating the bidding contractors' design solutions and price for designing and constructing the works. The same form of contract is often used as a basis for a negotiated contract.

Whilst it is possible for the design to be complete prior to construction, the form of contract envisages design by the contractor during the contract period. Insufficient thought to design by the contractor prior to acceptance of the contractor's proposals by the employer often leads to disputes as to what constitutes a variation to the employer's proposals and what ought to have been contemplated by the contractor as part of the original design. Comprehensive and detailed proposals by the employer can reduce the scope for such disputes.

2.6 Other forms of contract

Government forms of contract, such as GC/Works/1, are used extensively in the public sector. Amended versions exist for overseas projects. In the latest edition (Edition 3) much of the administrative work falls on the project manager appointed by the authority (the employer). There are contractual provisions for acceleration. Variations and amendments to the standard publication enable alternative methods of contracting to be used, such as design and build.

Other standard forms of contract issued by professional bodies are available and are worth considering as alternatives to some of the better known standard forms of contract.

In the civil engineering field, the ICE and FIDIC forms of contract are well established and are used in many parts of the world. The sixth edition of the ICE conditions of contract is now in use and it remains to be seen if this form of contract can maintain its almost universal recognition in the face of competition from new alternative forms of contract devised by leading experts in construction contracts.

The New Engineering Contract (NEC) (1991) reflects a substantial move to recognise, and cater for, the various forms of contract which have been discussed herein. It is based on a core contract with flexible alternatives allowing the employer to choose the appropriate version to suit his needs.

The ten document package consists of a core contract containing provisions which are universal to all versions. The various versions are:

- Document A – Conventional Contract with Activity Schedule;
- Document B – Conventional Contract with Bills of Quantities;
- Document C – Target Contract with Activity Schedule;
- Document D – Target Contract with Bills of Quantities;
- Document E – Cost Reimbursable Contract;
- Document F – Management Contract.

An engineering subcontract, guidance notes, flowcharts and other optional provisions pave the way for a better understanding of contracting methods and its use should be encouraged.

2.7 Special conditions and contract documents

In many building contracts, the standard conditions of contract are intended to stand on their own to be used without amendment. Where partial possession or sectional completion of the works is intended, some forms of contract may need special attention to enable these provisions to be incorporated. The Joint Contracts Tribunal have published several supplemental conditions of contract designed to be used with the appropriate standard forms of contract for these purposes. Failure on the part of professional advisers to give sufficient thought to these matters is a common cause of dispute which is often resolved against the interests of the employer.

The general rule of law is that a specially written document which forms part of a contract will take precedence over a standard document. Many construction contracts have gone to considerable lengths to negate this rule. The widely criticised provisions in clause 12(1) of JCT63 have survived and appear in JCT80:

> 'Clause 2.2.1 Nothing contained in the Contract Bills shall override or modify the application or interpretation of that which is contained in the Articles of Agreement, the Conditions or the Appendix.'

Similar provisions appear in many other JCT standard forms of contract (clause 4.1 of the Minor Works form; clause 2.2 With Contractor's Design and clause 1.3 of the Intermediate Form).

It is self evident, on the wording of the abovementioned provisions, that intended amendments appearing in other contract documents, such as the contract bills (of quantities) will be of no effect. It is also inappropriate to delete the relevant clause (such as clause 2.2.1 in JCT80). The deletion may cause *everything* in the other contract documents to override or modify the standard conditions, which may not be the intention without the most careful drafting of the other contract documents. If other provisions are intended to take precedence over the standard document, such provisions ought to be incorporated by additional clauses in *The Conditions* [of Contract]. Alternatively, supplemental conditions of contract may be used with an appropriate amendment to clause 2.2.1 of JCT80 (or the corresponding clause in other forms of contract) to give full effect to the supplemental conditions.

For the avoidance of doubt, the contract documents should be clearly specified. In the JCT forms of contract, the contract documents are described in the contract (for example, see clause 2.1 of JCT80). Sometimes other documents, such as exchanges of correspondence, are bound into the documentation with the intention of incorporating such documents into the contract. It is advisable to make the appropriate amendment in the conditions of contract giving full effect to other documents, setting out the order of priority in the case of ambiguity. If the latter is not done, it is likely that these other documents will take precedence (under the general rule). This may be acceptable if the entire contents of the other documents are to take precedence. However it is sometimes the case, after negotiation and clarification, that parts of the contents of such documents are not intended to apply. It is better practice to summarise any special provisions which may have been agreed in correspondence and incorporate such provisions in the contract. This will avoid the necessity to include correspondence in the documentation.

In civil engineering contracts, the contract documents are intended to be mutually explanatory of one another (clause 5 of the ICE fifth and sixth editions). The engineer is empowered to explain any ambiguities and make any necessary adjustment resulting therefrom. This is a potential cause of disputes, particularly where the drafting and editing of the contract documents (by the engineer who may be responsible for the ambiguities) are done without the necessary care.

In international contracts, the FIDIC conditions of contract provide for other documents to be incorporated by reference in the letter of acceptance or in the contract agreement. The order of priority of the documents forming the contract is specified (clause 5.2 of the fourth edition). This is a valuable feature which assists in dealing with ambiguities. Part II of the FIDIC conditions of contract contains the special conditions which take precedence over the standard conditions of contract. The use of this method encourages the standard of care necessary to draft clear and unambiguous contracts.

Other documents such as drawings, specifications and bills of quantities need careful attention to ensure that there are no ambiguities in, or between them. A common practice (to be discouraged) is the use of standard specifications or preambles which have not been edited to remove clauses which are not applicable to the work to be done. Every specification clause or preamble should be relevant to the work shown on the drawings. If it is decided to change the specification during the course of the project, then a new specification clause can be issued as part of a variation order. Some engineers and architects try to argue that contractors are required to carry out work which is not in the contract, at no extra cost, merely because it is mentioned in the specification.

Only the most careful editing of all of the documents forming the contract will minimise the exposure to claims arising out of ambiguities. Each contract should be treated as being unique and reliance on standard documents for all contracts should be discouraged in many instances.

3 Tender and Acceptance

3.1 Selection of tendering contractors: prequalification

Many mistakes and potential claims can be avoided if sufficient thought and planning is put into the pre-tender stage of a contract. A common mistake is to invite too many contractors, at the last possible minute, to submit a tender for a project. There have been cases of over twenty contractors being invited to bid for a project. In a recession, all, or most of the invitees will oblige. This process may provide the lowest possible tender figure. However, it does not guarantee the lowest final account and very often completion of the project on time (if the contractor survives the course) may be in doubt due to failure to resource the project properly. In a buoyant market, some contractors may submit *cover prices* (not a genuine tender, but one based on another tendering contractor's price and uplifted to ensure that it will not be successful). It has not been unknown for only one serious bid to be made alongside several *cover prices*. In such circumstances, the contractor submitting the serious bid usually discovers that fact and the tender price increases accordingly.

Substantial benefits can be gained by early selection of contractors who are willing to submit a bona fide tender and who are capable of carrying out the work. This can be done by carefully selecting potential contractors, giving them reasonable notice of the proposed tender and inviting them to indicate their willingness to submit a tender for the project. The invitation should contain sufficient information to enable the invitees to consider their ability to submit a tender and execute the work, such as:

- Date for issuance of complete tender documents;
- Date for receipt of tenders;
- Date of award of contract;
- Date for commencement of the work;
- Contract period;
- Form of contract (with or without bills of quantities);
- Liquidated and ascertained damages;
- Brief description of the project.

It should be made clear that any firms wishing to decline from submitting a tender would not prejudice their chances of being invited to tender for future work. Firms who accept the invitation should be given the opportunity to attend a preliminary meeting and view the drawings which are available.

If the above procedures are followed, the employer will be reasonably confident that he will receive serious bids from contractors. In the event of insufficient positive replies, the employer can widen his net to make enquiries of other firms. In addition, each contractor will be able to prepare for the necessary staff to be available and it can begin to make enquiries of potential subcontractors and suppliers.

In the case of large complex projects it may be desirable to invite contractors to pre-qualify to tender for the work. The procedures described above will be equally applicable to this process. However, in addition to providing the information mentioned hereinbefore, the employer will wish to find out more about the potential tenderers' capability. Prequalification enquiries should cover:

- Previous track record on similar projects;
- Proposed management structure and staff responsible for the project;
- Financial standing of the firm;
- Resources which can be made available for the project;
- Details of any *joint venture* if tenders are to be submitted in the name of more than one firm;
- Outline proposals for method of construction and programme.

In some circumstances it may be appropriate to include all of the matters described for management contracting in Chapter 2.

Prequalification inquiries should inform tenderers of the criteria to be used for selection. After receipt of prequalification documents from the invitees, a shortlist should be prepared according to the applicants' response, measured against the relevant criteria. This should be followed by interviews of the shortlisted firms and the final tender list should be drawn up as soon as possible so that all firms can be notified without delay.

With the advent of the Single European Act, a number of Directives issued by the European Commission have come into effect. The EC Public Procurement Directives cover work in the public sector, that is, work to be done by Contracting Authorities (government departments, local authorities, nationalised industries and private sector bodies receiving more than fifty per cent of their funding from government and all bodies governed by public law), the value of which exceeds specified thresholds (subject to review). The principal EC Directives relating to procurement are:

The Public Supplies Directive, *77/62/EEC* (amended 22 March 1988, *88/295/EEC*) – Governing supplies where the contract exceeds ECU 200 000 (£132 000). Exclusions include transport, distribution of drinking water, energy, telecommunications, contracts subject to secret or national security measures and certain contracts under international agreements;

The Public Works Directive, *71/305/EEC* (amended 18 July 1989, *89/440/EEC*) – Governing contracts for works exceeding ECU 5 million (£3.3 million);

The Excluded Sectors Directive, *90/531 EEC OJ* L2791/29 October 1990 – Governing public works and supply contracts in water, transport energy and telecommunications sectors where contracts exceed the thresholds given in the following table:

Sectors (Works)	– ECU 5 million (£3.3 million)
Water (Supplies)	– ECU 400 000 (£264 000)
Transport (Supplies)	– ECU 400 000 (£264 000)
Energy (Supplies)	– ECU 400 000 (£264 000)
Telecommunications (Supplies)	– ECU 600 000 (£396 000)

The Directives require prior indicative notice (planning approvals) and contract notice (details of the work which is the subject of the tender).

The criteria for selection of contractors include evidence of capability and a proven track record for five years, details of key staff, plant, equipment, labour and technical resources. References and financial information may be required. Failing to comply with certain laws, such as legal requirements to pay taxes and social security contributions may be grounds for disqualification.

3.2 Time allowed for tendering

It is unreasonable to allow only a few weeks to tender for a construction project of any reasonable size. Nevertheless, this is often the case. It is understandable that employers wish to start construction as soon as possible and it is this pressure which leads to insufficient time being allowed to enable tenderers to prepare a tender properly. Insufficient time often leads to numerous potential errors. A survey carried out in the United States in the 1970s indicated the following incidence of bid mistakes (*Anatomy of a Construction Project* by Kris Nielsen, *International Construction*, November 1980):

- Extension errors – 19 per cent (errors in multiplication to calculate quantities or price);
- Lack of knowledge of work required – 16 per cent (insufficient attention to all of the work involved);
- Lack of knowledge of contract administration requirements – 15 per cent (failure to identify risk or insufficient allowance for cost of administration);

- Under-estimating escalation – 12 per cent;
- Transposition errors – 10 per cent (transposing incorrect figures from one sheet or document to another);
- Poor pre-bid planning – 9 per cent;
- Poor resource planning – 9 per cent;
- Incorrect measurement of quantities – 8 per cent;
- Others – 2 per cent.

Given more time to tender for the work, the incidence and magnitude of errors ought to be reduced. A distinction must be drawn between mistakes in pricing by the contractor and mistakes on the face of the documents, such as incorrectly extending a rate for an item of work. It must be in the interests of both the employer and the contractor to avoid errors in the tender. A low bid due to one or more mistakes often causes the successful contractor to try every means to reduce costs and/or to pursue unmeritorious claims based on varying degrees of fiction.

However, it is not necessarily correct to assume that tenders will be higher if more time is allowed and errors are avoided. If competent contractors are given sufficient time to tender, they will be able to incorporate savings brought about by detailed studies into methods of construction, programming and procurement of plant and materials. Given that tenderers are in competition, some, if not all, of these savings will be passed on to the employer.

Many problems and mistakes can be avoided without delaying the date for receipt of tenders. Tenderers can be given more time if some of the tender documents are issued in advance of the entire set of tender documents. For example, drawings and sections of bills of quantities or specifications can be issued to tenderers before the preparation of the final tender bills is complete. A considerable part of a contractor's pre-tender planning and pricing will be based on the drawings. A detailed method statement will be prepared almost exclusively from drawings.

Tenderers often have to measure quantities of work from the drawings to determine plant size and other resources. This is the case even where bills of quantities are provided by the employer. Prices for special items are often obtained on the basis of the drawings. In many cases, tenderers may be able to establish, with reasonable accuracy, the cost of carrying out the works, before the final set of tender documents are issued. All that may remain to be done, during the relatively brief period allowed to submit the tender, is to thoroughly check the tender documents, obtain confirmation (or adjustment) of prices from subcontractors and suppliers, adjust costs where necessary, adjudicate on the final tender sum and compile the rates in the tender to arrive at the proposed tender sum.

A suggested timetable for the above is shown in Figure 3.1.

The EC Public Works Directive *89/440/EEC* lays down strict rules for tenders which are covered by the legislation. The open tendering procedure

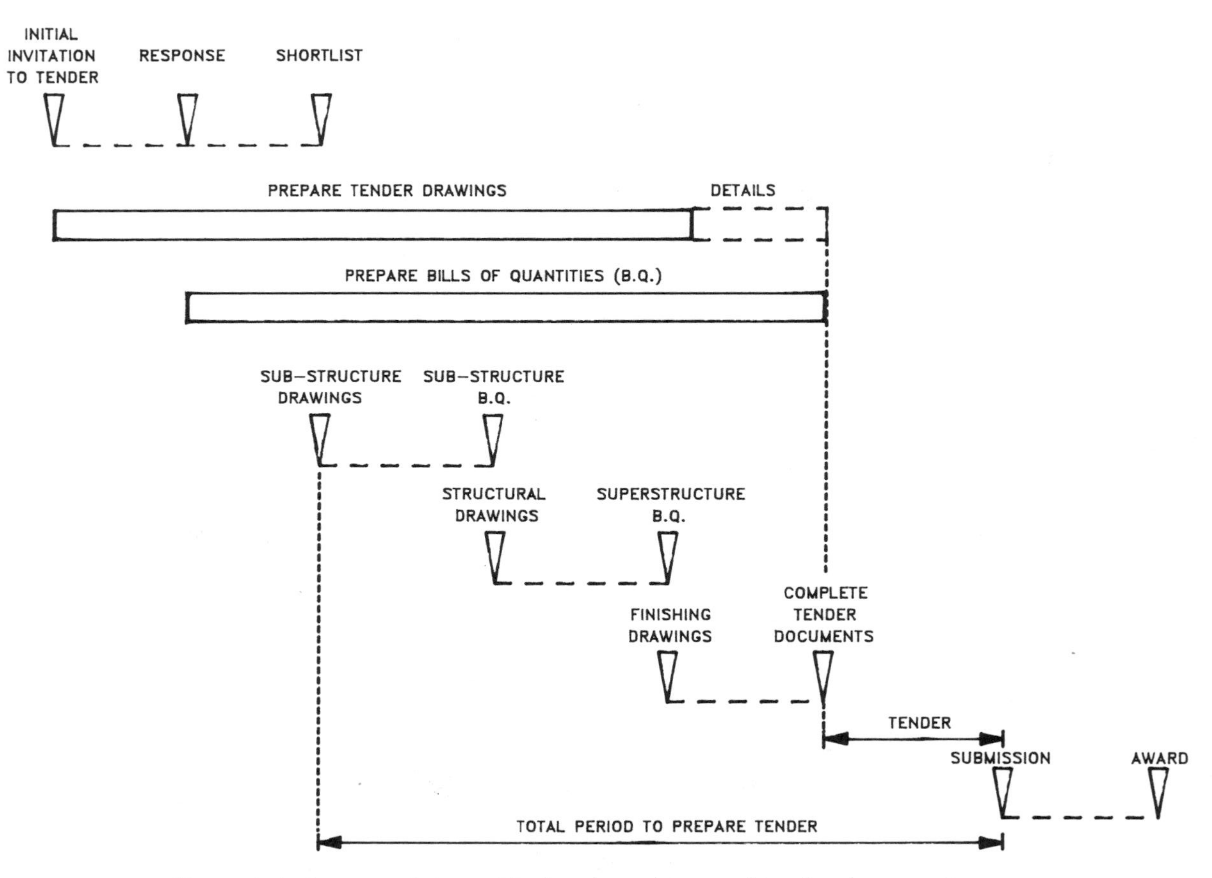

Figure 3.1 Suggested timetable for phased issue of tender documents

must allow a minimum of 52 days from dispatch of tender notice to receipt of tenders. The restricted (or selected) procedure must allow a minimum of 37 days from dispatch of tender notice to receipt of applications to tender and a minimum of 40 days from dispatch of written invitations to tender to receipt of tenders. The accelerated tender procedure may be permitted in some cases of emergency, in which case the periods may be reduced. Where no suitable tenders have been received during the normal tendering procedures, or where additional work is required in connection with an existing contract, direct negotiation with one or more contractors may be permitted.

3.3 Exploitation of poor tender documents by contractors

An increasing number of firms engage staff to scrutinise all of the tender documents to find ambiguities and other deficiencies that may be exploited to produce a lower tender and a potential claim for additional payment during the course of the project. It may be argued that all tenderers have the same opportunity to exploit such deficiencies, and the employer will end up paying no more, at the end of the day, than it would if the tender documents had contained no deficiencies.

This is far from the case. The successful contractor will often recover more, by way of claims, than it would if all of the costs had been included in the tender sum at competitive rates. In addition, extensions of time for completion of the works may flow from these deficiencies, whereas no additional time would result if there had been no deficiencies. Claims which arise out of innocent misinterpretation of the contractual intentions, or exploitation, where there is an ambiguity or deficiency, are often the most difficult to resolve amicably, since they reflect on the competence of the employer's professional team.

Contractors can assist in avoiding problems that arise out of ambiguities by notifying the employer's professional team of any ambiguity discovered at pre-tender stage. These ambiguities should then be rectified and brought to the attention of all tenderers prior to submission of tenders. If this is done, all tendering contractors will be tendering on an equal basis and the risk of exploitation will be minimised.

The employer's professional team should take care when evaluating tenders so that any obvious pricing anomaly (between tenderers) is reviewed with the tenderers to establish the reason for it.

3.4 Preparing the estimate: adjudication: the tender

The estimator's task is to accurately calculate the cost of carrying out the works and to apportion the cost to the various elements (or items in a bill of quantities) of the job. In order to do this he may have to rely on several other departments, or individuals, in the company. The cost of carrying out the works is very much determined by the method of construction and the programme for the project. The method of construction will determine the type of plant to be used and the productivity to be expected. The programme will determine the cost of time related items such as external scaffolding, tower cranes and hoists. The amount of work to be subcontracted may determine the number of supervisory staff and the cost of attendance on each subcontractor. Compiling the estimate is a completely separate task from tendering. The estimator should not make decisions or allowances which are influenced by external market forces or post-contractual matters such as *front loading* the rates (increasing the rates for work executed early in order to improve cash flow). He may, however, advise management on such matters.

Once the estimate has been compiled and the cost of executing the work has been established, management will consider external factors such as the competition and the probable successful tender sum. The existing workload of the company and the requirement to obtain further work will also be considered, as well the assessment of risk, staff resources, profit and possible savings in cost which can be made. This process is known as *adjudication.* After due consideration of all of these factors, the estimate will be converted into the tender for the works. The estimator will then make all of the necessary adjustments to the rates in accordance with the decisions of management. The form of tender will then be completed and submitted. In times of recession, the tender sum may be less than the estimate of cost for executing the works.

A typical estimating and tendering process is shown in Figure 3.2.

3.5 Qualified tenders

Some public corporations and government departments are bound by rules which preclude the acceptance of a qualified tender unless all tendering contractors are allowed to modify their tenders to incorporate the same terms and conditions. Some are prohibited from considering a qualified tender at all. Apart from the above considerations, are there any reasonable grounds to qualify a tender?

Tendering contractors may suspect a risk if certain representations are made by the employer such as the availability of materials provided by the employer or as to the ground conditions. Careful examination of the proposed contract conditions or knowledge of the general law may render a qualification unnecessary, in which case none should be made as it detracts from what would otherwise be a complying tender. On the other hand, the proposed contract terms may be particularly onerous. The tendering con-

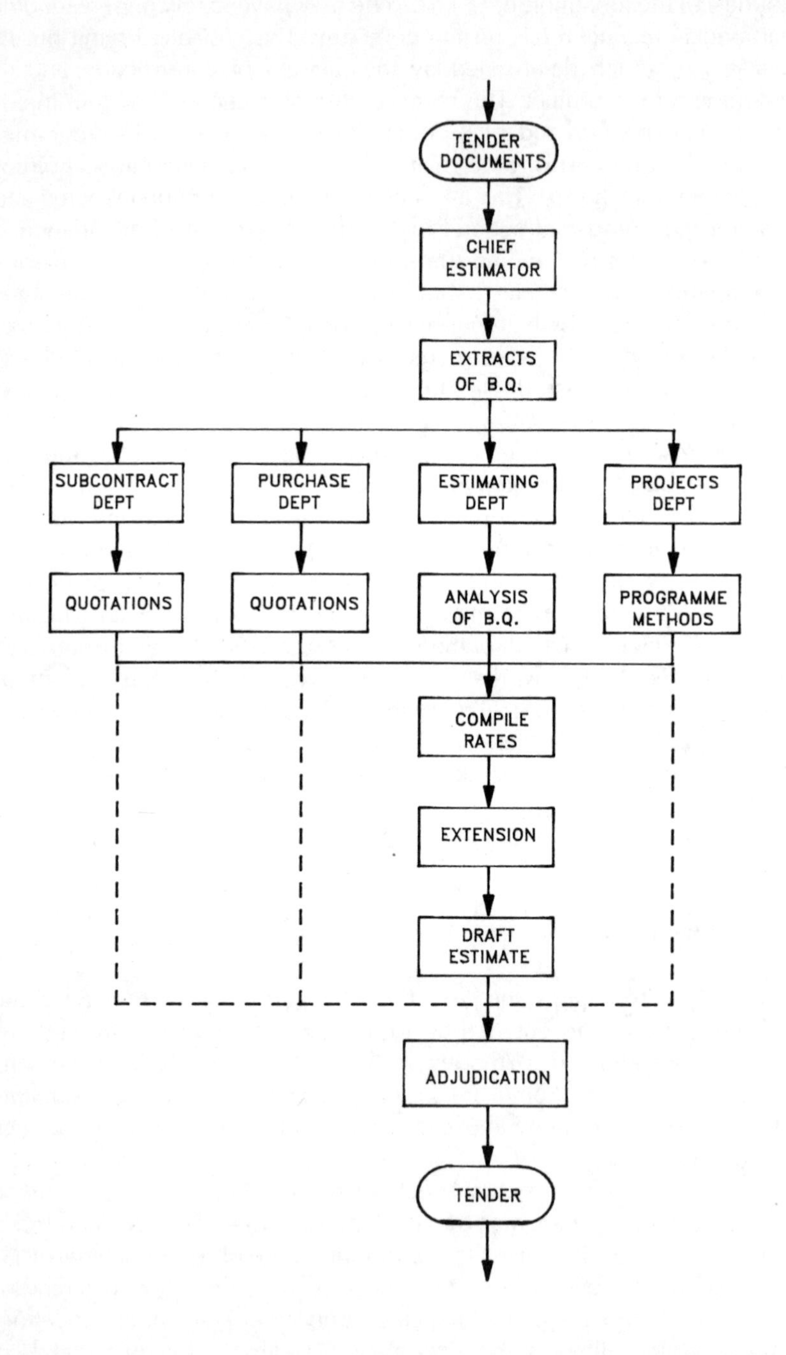

Figure 3.2 Typical estimating and tendering process

tractor then has the option of pricing the onerous terms (which may not be possible without an element of gambling) or qualifying the tender in order to have the onerous terms modified or removed.

From a practical point of view, if the employer is properly advised, it may be sensible to invite a complying tender and an alternative tender incorporating certain changes which may be proposed by the tenderer. It could be a condition of tender that all proposed changes should be notified several days before the date for receipt of tenders, with the proviso that all tendering contractors will be informed of the proposed changes. If that is done, all tenderers will have the opportunity to submit an alternative bid incorporating those changes that they saw fit to adopt. If each adopted change was required to be priced individually as an omission from, or addition to, the complying bid, it would assist in evaluation of tenders and there would be no delay in making an award. If qualifications are permitted without prior notification on the date for receipt of tenders, there will almost certainly be delay caused by evaluation and possible re-tendering. By that time all of the tendering contractors will have a reasonable idea of the lowest tender, in which case there is room to make other adjustments in order to make the revised tender more competitive.

If a qualification is made to a tender, it is important to ensure that it is couched in terms which make it a condition and that it is incorporated in the contract. If extra costs are involved, the contract terms should clearly state how these extra costs are to be added to the contract price and in what circumstances. Qualifications contained in the tender, or in a letter attached to the tender, will only be effective if the tender (or letter) is a contract document, *Davis Contractors Limited* v. *Fareham U.D.C.* [1956] AC 696. Alternatively, the qualification should become a contract term by modifying the conditions of contract.

3.6 Tender programme

The preparation of a tender programme is essential. It is an important aid to the contractor when assessing cost and resources and to the employer when evaluating the tender. In many cases a simple bar chart will suffice. However, for complex projects, a detailed programme showing the logic and restraints is required. The programme should be realistic. All too often, the programme which is submitted is no more than a tool to form the basis of potential claims which may arise. The contractor is usually required to complete the project *on or before the date for completion*. Some contractors deliberately show early completion. If this is possible without a disproportionate increase in cost it is often in the interests of both parties to agree an earlier completion date. Problems can occur if the contractor's tender is accepted and completion is shown, on the programme, at an earlier date than the contractual date for completion (*Glenlion Construction Ltd* v. *The Guinness Trust* (1987) 39 BLR 89 – see Chapter 5).

The tender programme will not usually be a contract document, but it is often relied upon when formulating claims. For this reason it must be a document which is a genuine reflection of the contractor's intention and evidence to support this may be necessary. Estimated productivity, logic, proposed plant and methods are some of the matters which may have to be considered in detail to justify the contractor's programme.

Considerable areas of doubt may exist in any programme which relies upon prime cost and provisional sums for important elements of the project. The tendering contractor is required to allow for the completion of all of the work by the contractual completion date. It is good practice to indicate, on the programme, the sequence and duration of work to be done in respect of each and every prime cost and provisional sum. Ordering periods, relationship to other work and durations of the prime cost or provisional work, which have been assumed, should be clearly indicated. Wherever possible, the employer should inform all tendering contractors of proposed, or potential, nominated subcontractors and suppliers so that the programme requirements can be based on realistic information obtained from them. Any additional information regarding provisional items should be given to the tendering contractors so that the element of guesswork is reduced or minimised.

3.7 Evaluation criteria

Some public bodies are prohibited from accepting tenders on the basis of any other criteria than the lowest price (errors excepted). The lowest price does not guarantee the lowest final account, and a detailed analysis of tenders can sometimes indicate a possible exposure to a higher price than the tender sum.

Save where tenders are very close, the acceptance of the lowest tender may not be in the employer's best interests. A very low tender should not normally be accepted without first discussing every contentious matter with the tenderer. Errors should be dealt with in accordance with one of the codes of practice (which should be notified to tenderers prior to submission of tenders).

However, for some projects, price alone may not be the criteria which determines the best bid. The tender programme may indicate to what extent the tenderer has appreciated the complexity of the design. Proposed methods may indicate to what extent the tenderer has appreciated the details and co-ordination of services. It is essential that the employer sets out the criteria, giving each a standard, or yardstick, by which tenders are evaluated. Tendering contractors should be made aware of the evaluation criteria to be used so that the tender can be prepared accordingly.

Evaluation can be assisted if tenderers are required to submit additional information in support of the tender. This may include:

- Breakdown of major items into labour, plant, materials, overheads and profit;
- Breakdown of costs related to time, volume, method and event;
- Cash flow forecast.

Rates inserted in schedules, or bills of quantities, by the tenderers should be examined and compared to ensure that there are no obvious and significant departures from what is considered to be reasonable. Suspect rates may be due to ambiguous descriptions, mistake as to quality, failure to allow for materials or other causes. Inconsistencies in rates (between sections of bills of quantities) should be adjusted by agreement if it is appropriate.

Final selection should not take place before interview with the tenderer. Key staff proposed by the tenderer should attend the interview and all important matters should discussed in detail to ensure that there are no problem areas that cannot be resolved.

The criteria for the award of contracts laid down in the EC Directives are lowest price or most economically advantageous tender. In most cases, lowest price will be the deciding factor. If the latter is to be adopted, the contracting authority is required to advertise the fact giving a list (and if possible, the order of priority) of the criteria to be used in evaluating tenders. Matters such as completion periods (which may be a competitive element), maintenance costs, costs in use and technical specifications may be used for evaluation purposes.

3.8 Rejection: acceptance: letters of intent

In the normal course of events (and subject to certain criteria laid down in the EC Directives), there will be no problem if a tender is rejected. However, in the event that a tenderer has been required to do a substantial amount of preparatory work which is outside the scope of that which is normally required, the tenderer may be entitled to payment. In the case of *William Lacey (Hounslow) Ltd* v. *Davis* [1957] 2 All ER 712, it was held that there was no distinction between work done which was intended to be paid for under a contract erroneously believed to exist and work done which was intended to be paid for out of proceeds of a contract which both parties erroneously believed was about to be made. Such work was not done gratuitously and a reasonable price must be paid for it. The same principle was applied in *Marsden Construction Co Ltd* v. *Kigass Ltd* (1989) 15 ConLR 116.

The EC Directives provide that tenders may not be rejected because they appear to be too low, without allowing the tenderer to give an explanation.

In *Fratelli Costanzospa SpA* v. *Comune di Milano (Municipality of Milan)* [1990] 3 CMLR 239, an unsuccessful tenderer commenced proceedings against the Municipality on the grounds that his tender had been rejected pursuant to the Municipality's formula which automatically rejected all tenders which were more than ten per cent lower than the average of all tenders. It was held that the tenderer had the right to seek enforcement of the Directive.

The Directives also forbid rejection on the grounds that the tender is based on equivalent alternative specifications which meet ISO standards. In *Commission of the European Communities* v. *Ireland* (1988) 44 BLR 1, an Irish company complained that its tender was rejected because the Spanish products offered by the tenderer did not comply with Irish standards specified in the tender documents. The Spanish products complied with ISO standards and it was held that the contracting authority (Dundalk Urban District Council) had failed to comply with Article 30 of the Treaty of Rome by excluding products of equivalent ISO standards. It should be noted that this particular contract was excluded under the threshold provisions of the Public Works Directive, but it was not exempt from the general provisions of the Treaty of Rome for non-discriminatory technical specifications.

Errors in tenders should not normally be cause for rejection. Where errors in the tender are discovered and dealt with in accordance with the relevant codes of practice, many potential problems can be avoided. In any event, if the employer discovers an error in the tender before acceptance, and the tender is accepted without adjustment, the contractor will not be bound by the error: *McMaster University* v. *Wilchar Construction Ltd* (1971) 22 DLR (3d) 9 – High Court of Ontario.

Tenderers are often asked to keep their tenders open for acceptance for a specified period. This does not prevent the tenderer from withdrawing his tender at any time. Tenderers may be bound by their tenders if there is consideration. The amount of consideration may only be nominal. Alternatively, a *Bid Bond* may be required by the employer. Once the employer has unconditionally accepted a tender within the time for acceptance of tenders (or within a reasonable time if there is no specified time) and provided that the tender has not been withdrawn, there is a binding contract.

Post-tender negotiations often take place, particularly in the private sector. Public tenders are less likely to be subject to negotiation. Current EC law does not cover post-tender negotiations. However, the Council of Ministers have issued a statement on this matter:

> 'The Council and the Commission state that in open or restrictive procedures all negotiations with candidates or tenderers on fundamental aspects of contracts, variations in which are likely to distort competition, and in particular on prices, shall be ruled out; however, discussions with candidates or tenderers may be held but only for the purposes of clarifying

or supplementing the content of their tenders or the requirements of the contracting authorities and providing this does not involve discrimination.' *Public Procurement Directives,* conference paper by Robert Falkner, 10 December 1990.

It is not unusual for acceptance to be conditional, usually by way of a letter of intent. Care should be taken by the employer when drafting a letter of intent. Equally, the contractor should carefully consider the terms of a letter of intent in order to understand fully to what extent he has been authorised to proceed and how payment for work done will be established. Matters to be addressed when drafting a letter of intent should include:

- Detailed instructions clearly describing the work which is to proceed, distinguishing between design, ordering, taking delivery and execution of work;
- Full compliance with the tender documents so far as they apply to matters for which authority to proceed has been given;
- Terms of payment to be made in respect of the matters for which authority to proceed has been given;
- Provision for termination of contractor's rights to proceed pursuant to the letter of intent and the employer's liability for payment in the event of termination;
- Provision for cancellation of orders placed pursuant to the letter of intent and the employer's option to pay cancellation charges or to take delivery of goods ordered;
- Care of, and responsibility for, work and materials including insurance;
- Goods and materials to be vested in the employer;
- Provision to terminate the terms of the letter of intent in the event of award of the contract and provisions to credit payments made under the letter of intent against certificates issued under the contract;
- Provision for settling disputes (usually retaining the same provisions as the proposed contract).

It is important that the letter of intent should make it clear that it is not acceptance of the contractor's tender. It should, however, make it clear that the employer has the option to accept the contractor's tender.

Even the most carefully prepared letter of intent may have its problems. In *British Steel Corporation* v. *Cleveland Bridge Engineering Co Ltd* (1981) 24 BLR 94, the courts had to consider whether, or not, a contract had been created by a letter of intent. It was considered that each case must depend on the particular circumstances. However, it was decided that if a party acted on a request in a letter of intent and was simply claiming payment, it did not matter if a contract was not created as payment could be based on *quantum meruit.*

In *C.J.Sims Ltd* v. *Shaftesbury Plc* (1991) QBD; 8-CLD-03–10, the court had to consider the payment terms of a letter of intent. The terms provided for reimbursement of reasonable costs, including loss of profit and contribution of overheads 'all of which must be substantiated in full to the reasonable satisfaction of our quantity surveyor'.

At first glance it would appear that the above terms were reasonable commercial requirements for payment. The employer successfully argued that it was a *condition precedent* to any payment being made to the contractor that the costs should be *substantiated in full* and *to the satisfaction of the quantity surveyor.* The judge was not disposed to the view that the contractor should be paid something on account pending full substantiation (which, with respect, is what would normally be expected).

A potential disaster area exists when contracts proceed on the basis of protracted correspondence and exchanges of letters, all of which contain elements of change to previous documents and there is no clear definition of the terms agreed between the parties. In *Mathind Ltd* v. *E. Turner & Sons Ltd* [1986] 23 ConLR 16, the contract was intended to be JCT63. Exchanges of correspondence and an addendum bill of quantities dealt with phased handover. The works proceeded but the contract was never signed. Disputes arose over phased completion dates and liquidated damages. The court had to consider when and how the contract was made. In doing so it came to the conclusion that both parties had agreed to phased completion. As no contract had been signed the contractor could not rely on the words in clause 12(1) of JCT63 which prohibited modification to the standard printed form in the contract bills. (It should be noted that in *M. J. Gleeson (Contractors) Ltd* v. *London Borough of Hillingdon* (1970) 215 EG 165, provisions for phased completion were contained in the contract bills. The provisions were held to be ineffective on the grounds that the contract stipulated that nothing contained in the contract bills should override or modify in any way the contract conditions.)

It is not uncommon to agree to change the conditions, or specification or details, in the tender documents, prior to signing the contract. Failure to amend the contract documents to reflect the change may mean that the change, when made, is a variation to the contract despite the fact that the parties had agreed to the change prior to signing the contract. In *H. Fairweather & Co Ltd* v. *London Borough of Wandsworth* (1987) 39 BLR 106, the contract was signed after both parties had agreed that the specified *Clifton* bricks would not be used and that *Funton* bricks would be substituted therefor. There was delay in delivery of *Funton* bricks. The contractor claimed that the delay arose out of a variation and claimed an extension of time under clause 23(e) and loss and expense under clause 11(6) of JCT63. The architect granted an extension of time under clause 23(j)(ii) for unforeseen shortages of materials, and refused a claim for loss and expense. It was held that the substitution was a variation.

In view of the above, it is essential that all agreed changes to the tender documents should be reflected in the contract to be signed by the parties.

Any agreed change which would otherwise constitute a variation should be reflected in revised contract bills. If any change affects the completion dates previously mentioned in the tender documents, the appropriate adjustment should be made in the contract documents prior to signature. If necessary, the tender (or contract) programme should be revised.

Finally, with the exception of essential key dates, it may be fatal to incorporate the contractor's programme as a contract document. Acceptance of a tender may be on the basis of the contractor's programme, but its use as a contract document can cause considerable problems. This aspect will be dealt with in Chapter 4.

4 Monitoring Delay and Disruption Claims: Prevention

4.1 Contracts administration

All forms of contract contain express or implied duties and obligations to be performed by the employer (or his agents) and the contractor. Contracts do not usually set out in detail how these duties and obligations should be performed. It is self evident that the employer must give access to the site and provide information in sufficient time to enable the contractor to carry out the works by the due completion date. The contractor must give reasonable notice of delay or of any claim and the architect, or engineer, must decide and make extensions of time or certify additional payment.

Whatever the form of contract, it is important that all parties co-operate with each other in order to ensure that each is provided with sufficient information to enable them to carry out their respective duties and obligations. Too often, contractors believe that they have complied with their contractual obligations by giving notice of delay and very brief information (if at all) to support their contention that they are entitled to more time and/or money. It is not unusual for contractors to complain that no extension (or insufficient extension) of time has been granted by the architect or engineer. These complaints sometimes persist several years after the contract has been completed when the first pleadings are being prepared for arbitration. Even at this stage some contractors are unable to show what period of delay occurred and its effect on the progress of the works. Criticism of the architect, or engineer, for failing to make an extension which satisfies the contractor is hardly justified (provided of course that an honest attempt was made to assess the effects of the delay) if the contractor, himself, cannot illustrate the effects of the delay.

These problems can be avoided if all parties examine the contract terms to establish their express duties and obligations and what procedures need to be adopted in order to ensure that these duties and obligations can be performed in accordance with the contract.

Whatever procedures are to be adopted, they should not become a costly and time consuming burden so that resources are diverted from the main objectives of any building and engineering contract – to design and build the works.

4.2 Possession of site: commencement

Before award of the contract, the employer and the contractor should agree on the period of notice to commence in order to allow for mobilisation and the taking of records and photographs showing the condition of access and of the site prior to possession by the contractor. Any restriction or limitation on the free use of the site should be recorded and the effects (if any) on programme or cost should be established as soon as possible. Contractual provisions which envisage possession of the site being given to the contractor within a short period (for example, seven days) should be avoided if possible. Consideration should be given to allowing the contractor to mobilise and set out even if there are outstanding approvals which are essential to commence construction of the permanent works. Early access to the site should be distinguished from the contractual date which is the commencement of the period for completion of the works.

4.3 Pre-commencement meeting

Prior to possession of the site (if practicable before award of the contract) the parties and their professional advisers should convene a meeting to discuss and record certain important matters. These should include:

- The role and authority of each member of staff participating in the project;
- Where the contract provides for delegating powers to other persons, these powers should be clearly established;
- Status of the programme, key dates for information, periods for approval, long delivery periods and special problems;
- Requirements for named, nominated and selected domestic subcontractors;
- Works or materials to be provided by the employer;
- Procedures for interim valuations and certificates;
- Procedures for measurement, records, notices, particulars to be provided and response;
- Procedures for monitoring the progress of the works, photographs, video, progress records and updating programme.

It is important that the representatives of both parties understand the need to recognise potential delays and to acknowledge that they may lead to claims from the contractor and subcontractors. Whatever procedures are adopted at this initial meeting, they should include measures to avoid or minimise delay by regular monitoring of design and detailing so that the construction of the works will not be affected by late issuance of essential information.

4.4 Regular progress meetings

Meetings should be kept to a minimum, but should be sufficient to satisfy the needs of the project. Each meeting, or series of meetings, should be designed to suit specific objectives, have the right persons present and take place at the right time or at sensible intervals.

Three categories of person should attend; those who can inform; those who can advise and those who can (and are authorised to) decide on the issues and delegate action.

The most important features of successful meetings are:

- The correct agenda;
- Accurate records of the meeting;
- Decisions taken;
- Identify responsibility for action;
- Record of action taken (or outstanding) in respect of previous matters;
- Accurate forecasts or projections;
- Prompt distribution of minutes.

Where minutes of meeting are inaccurate, or where there are important omissions, it is essential that these are brought to the attention of the attendees and the necessary corrections made. Matters which require immediate attention should be dealt with in writing before the next meeting. Failure to follow these procedures causes major difficulties when trying to establish facts several years after the event. It is not unusual, when interviewing material witnesses in preparation for arbitration, to be told that the minutes of meetings did not record what was agreed. Even if it is possible to verify such allegations, it is sometimes difficult to reconstruct the history of events.

Records of meetings can often mislead investigators searching to establish causes of delay several years after the event. A common practice adopted by contractors is to table a long list of alleged outstanding information at each meeting. Many items reappear week after week and month after month. It is often difficult to distinguish between information requested far in advance of being required and information which was essential but which was neglected by the architect or engineer. Each alleged outstanding item should be addressed during the meeting, or by written response before the next meeting, giving the status and anticipated date of issue, together with a note indicating the programme and progress of any work which may be affected by the outstanding information.

The agreed minutes including any amendments should be signed by authorised representatives as a true record of the meeting.

4.5 Instructions and drawing issues

Many instructions and drawing issues are of an explanatory nature to enable the contractor to construct the original works. Late issuance of information will lead to claims for delay and/or disruption. The designer must be able to understand the contractor's programme and make allowance for shop drawings (if applicable), obtaining quotations, ordering and delivery. The designer should not rely solely on the contractor's requests for information (usually the contract does not place an obligation on the contractor to make any such requests). It is essential to have regular meetings to determine when information is required in order to meet the programme or to prevent delay.

Few construction contracts proceed without changes of some kind. Revised drawings should clearly indicate the revisions so that the contractor can identify appropriate action without searching to find each revision. Such drawings should be accompanied by a variation order/instruction to facilitate cost monitoring and control as well as indicating a possible review of the effects on programme.

Some architects and engineers issue drawings under cover of instructions, letters, transmittal sheets and other forms, without distinguishing between explanatory details and changes to the original design. This practice does not facilitate control and often contributes to failure, by the contractor, to give notice of delay, or extra cost at the earliest possible time.

4.6 Site instructions: verbal instructions

There is an increasing tendency to design and detail the works as they proceed at site level. This indicates lack of knowledge of design and construction detailing. Projects which end in protracted disputes have often suffered from an unusually high proportion of design and detailing by way of verbal instructions and hand drawn sketches issued by the designer's site representative during a regular 'walkabout' on site. It is not unusual, when investigating causes of delay and disruption, to discover numerous references in minutes of meetings to the effect that the contractor was instructed to proceed in accordance with a sample, or method, agreed on site. Records of what was agreed are often difficult, or impossible, to find. Interviewing site staff months, or years, after the event sometimes assists in this exercise at considerable expense. A dimensioned sketch and/or photograph at the time of the agreement would avoid any misunderstanding about what was required and built.

Site instructions and verbal instructions should be used in an emergency only and not as a method of designing the works. Where verbal instructions are given, the architect, or engineer, should take the initiative in making sure that they are confirmed (whether or not there is provision in the contract for confirmation by the contractor which would give effect to such instructions).

4.7 Form of instructions

Most contracts do not require an instruction, or variation order, to be in a particular form. A written site instruction, provided that it is issued by a person with the contractual authority to give instructions is, for all the purposes of the contract, an instruction authorising the contractor to proceed. It is effective without the need for a standard form of instruction to confirm its contractual effect. Likewise, a drawing issued by an authorised person is an instruction in its own right, regardless of the form of the accompanying covering instrument (or if there is no accompanying covering instrument).

Without proper agreed procedures and consistency for the issuance of instructions, whether they are explanatory or variations, there is an increased probability that monitoring and control of cost and delay will be ineffective. Very often, the full effects of all of the instructions issued during the course of the project do not come to light until the final account is on the table and the contractor is reconstructing (with hindsight) the history of events in order to resist a claim for liquidated damages levied for late completion.

4.8 Programme and progress

With the exception of some of the more recent engineering forms of contract, and the third edition of GC/Works/1, most standard forms of contract do not place sufficient emphasis on a construction programme. It is sometimes not even mentioned or required. Having regard to the sums of money spent on some modern projects and what might turn on events which affect the contractor's programme and progress, it is essential that a realistic programme showing how the contractor intends to construct the works should be available at the outset (see Chapter 3).

There may be problems if the contractor's programme becomes a contract document as failure to follow it in every detail may be a breach of contract. The contractor's obligations are normally to complete the works (or sections of the works) by given dates. Departures from the programme will be of no significance so far as the employer's remedies for performance are concerned. If there are good reasons for introducing key dates (for example, to facilitate installation of plant and equipment by the employer or specialists), these can be incorporated as contractual requirements, with appropriate remedies in the event of the contractor's failure to meet these key dates.

Another problem (when programmes become contractual documents) arises in the event of it being impossible to carry out the work in accordance with the programme. In the case of *Yorkshire Water Authority* v. *Sir Alfred McAlpine and Son (Northern) Ltd* (1985) 32 BLR 114, the contractor's programme and method statement became contract documents. The method statement, which was the contractor's own chosen method of working, provided for an outlet to a culvert to be constructed by proceeding upstream. The contract obliged McAlpine to execute the works 'in all respects in accordance with the contract documents'. It was found that this method was impossible and McAlpine successfully argued that it was entitled to a variation order to enable it to carry out the work. (It should be noted that the contract was based on the ICE conditions which provided, in clause 13(1), for the contractor to be relieved of its obligations to carry out work which is physically impossible.)

Having commenced work on the basis of a realistic programme, any significant departures from it should be monitored. Once delay has occurred which affects any important activities, it is essential that the effects of the delay are monitored, and that the programme is immediately updated to show the effects of the delay. If actual progress is monitored against a programme which is no longer valid, it is difficult, or even impossible, to establish the effects of a particular delaying matter on the overall programme and completion date. All progress, and delays, should be monitored against a programme which represents the contractor's proposed 'programme of the day', that is, a programme which has been revised to take account of all previous delays. As delays occur, these affect critical and non-critical activities. If regular updating is not done, the critical path may change, making the assessment of the effects of further delays a matter of guesswork. An example of how a critical path may change is given in Figure 4.1. In practice, this is no simple matter, and on contracts which have numerous, and often, continuing delays, it can only be achieved by additional staff and the use of various software and computers. It can be a costly exercise, and periodic updating may be a compromise which achieves reasonable results at an acceptable cost.

4.9 Notice: records and particulars

Many delay claims by contractors fail due to lack of notice and/or failure to justify any (or sufficient) extension of time, or additional payment, due to lack of records. No truer comment has been made than that made by Max W. Abrahamson in his book *Engineering Law and the I.C.E Contracts, fourth edition* at page 443; quote: 'A party to a dispute, particularly if there is arbitration, will learn three lessons (often too late):the importance of records, the importance of records and the importance of records.'

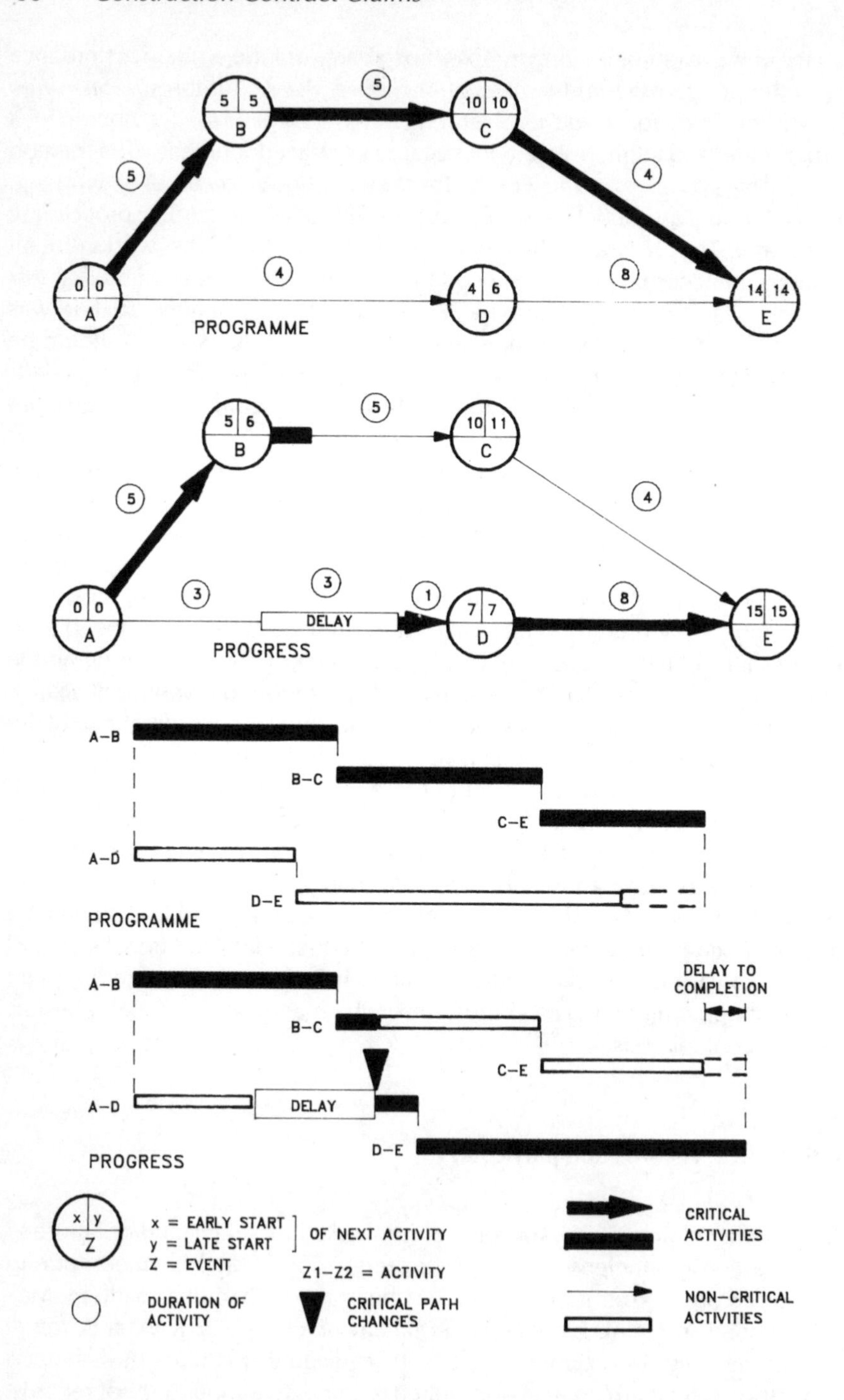

Figure 4.1 Example of change in critical path caused by delay

Whether, or not, there are contractual requirements to give notice of delay, or extra payment, contractors must, if they are to maximise the relief, or compensation, within the contractual remedies, give written notice of the delay or circumstances giving rise to the claim. Where the contractual provisions are stringent (and particularly where they are *conditions precedent*), contractors should ensure that each, and every, member of staff be made aware of these requirements and that each knows what role to play within contractual procedures designed to manage all delay and disruption claims. Where the contractor's staff have a good working relationship with the employer's staff, all notices should be clearly set out, identifying the contractual provisions under which the notice is being given, together with sufficient information to enable the recipient to be aware of the actual, or likely, effects of the matters in respect of which the notice is being given. In the unfortunate (and sadly, too frequent) cases where notice of any kind, no matter how well justified, produces a hostile reaction and continuous allegations aimed at 'muddying the waters', there may be some justification in couching the terms of any notice so that it is almost disguised. If this approach must be adopted, the significance of the notice must be capable of being understood in the light of other documents and the surrounding circumstances.

Having given notice, the contractor should keep contemporary records in order to illustrate the effects of the events, or circumstances, for which notice has been given. The recipient (the architect, or engineer) should also keep contemporary records. It is good practice to agree what records should be kept, to jointly monitor events and to agree facts during the progress of the works. Many contracts now contain express provisions for keeping records. Failure to agree facts is often caused by attempting, at the same time, to establish liability and entitlement. If both parties address their minds solely to agreeing *facts as facts*, leaving liability and entitlement for another day, agreement may be more readily achieved.

The most common records which ought to be kept are:

- Master/Detailed Programme and all updates with reasons for each update (preferably showing delays to each activity);
- Adverse weather conditions, including high winds and abnormal temperatures;
- Progress Schedule indicating actual progress compared with each revision of the programme;
- Schedule of resources to comply with the original and each revision of the programme;
- Records of actual resources used based on progress;
- Cash flow forecast based on the original and each revision of the programme;
- Records of actual cash flow;

- Schedule of anticipated plant output;
- Records of actual plant output on key activities;
- Records of plant standing and/or uneconomically employed (with reasons);
- Schedule of anticipated productivity for various activities;
- Records of actual productivity on key activities;
- Schedule of anticipated overtime (and the costs thereof) in order to comply with the original and each revision of the programme;
- Records of actual overtime worked and the costs thereof;
- Progress photographs and (where appropriate) photographs of work to be covered up;
- Where appropriate, video records showing sequence and method of working;
- Drawing register with dates of each revision and notes of amendments;
- Site diaries and dairies of key staff;
- Minutes of meetings and notes kept at meetings;
- Cost and value of work executed each month (for the project);
- Cost and value of work executed each month for all projects (company turnover);
- Allowance for overheads and profit in the tender sum;
- Cost of head office overheads each month (quarterly or yearly if not possible on a monthly basis);
- Profit (or loss) made by the company for each accounting period.

Many contractors do not have the management information systems or procedures to keep all of these records. However, many of them are capable of being kept on site with the minimum of extra effort. It is important to specify what records should be kept by different members of staff. For example, the contents of the diary, and records kept by the project manager will be different from those kept by a section foreman. Company policy should lay down procedures and guidelines so that there is the minimum of duplication (save where it is essential for verification) and that there are no gaps in the information to be collected.

On the employer's side of the fence, the architect, engineer, clerk of works and other staff should know what records they should each keep. If they are not kept jointly with the contractor, they should be agreed wherever possible. Keeping records for the purposes of defeating a claim in an arbitration may appear to be good practice, but it is more sensible to use them to settle contentious issues at the time so as to avoid costly disputes. In addition, if the contractor is aware that his grounds for a claim are doubtful (having regard to better records kept by the employer's professional team), it is more likely that the claim will be dropped and he will make an effort to get on with the job and possibly make up some lost time.

The employer's professional team should keep additional records to monitor delays by the contractor and delays for which no additional payment is payable.

Whatever records are kept, they are likely to be invaluable in the preparation of particulars in support of a claim. It should be remembered that particulars should, in addition to supporting the claim, be persuasive. It is all very well merely submitting all relevant records as particulars without some argument and illustration to set out the contractor's case and the entitlement sought, on the basis that it is the architect, or engineer, who is responsible for assessing the claim. Once the architect, or engineer, has made their assessment, it is sometimes difficult to persuade them to change their minds. Their assessment may be insufficient because they did not appreciate the effects of some delays on the method, sequence or timing of an operation, or because they did not recognise the significance of some of the records submitted. Naturally, they may be reluctant to admit this fact, particularly if it will bring to light their inexperience, or emphasise that the delay was due to their own incompetence. Good particulars should, in addition to providing supporting records, illustrate the effects of the events, or circumstances giving rise to the claim. To this end, the contractor is well advised to provide details and diagrams indicating:

- What ought to have occurred if there had been no delaying event, or circumstance;
- What actually occurred as a result of the delaying event, or circumstance;
- Analysis of facts, calculations, explanations and arguments to show how the delaying event, or circumstance, was responsible for the change in the method and/or programme.

4.10 Delays after the contract completion date

The best advice that can be given to any employer is not to cause any delay after the contractual completion date (extended, if applicable) has passed and when the contractor is in culpable delay. Very few contracts deal with delays by the employer after the completion date, and in many cases, once such a delay has occurred, the time for completion is no longer applicable and the contractor is allowed a reasonable time for completion of the works. Even where the contract does provide machinery for extending the date for completion in the event of such delays, there are few guidelines as to how the extension should be dealt with, and the effects on the employer's rights to liquidated damages. The Singapore Architects Standard Form of Contract contains very detailed provisions in clause 24 (see Figure 4.2). In this form of contract, it is intended that the employer may recover liquidated damages during a period of culpable delay by the contractor (even if a concurrent qualifying delay should occur during the period of culpable delay). Only if the contractor is not himself in delay is it intended that the employer's rights to recover liquidated damages be suspended during a further delay caused by a qualifying event or circumstance. However, with the greatest respect to the distinguished author of these provisions, they are unduly complicated, and they are likely to fail to protect the employer's rights to liquidated damages if the delay which occurs (after the completion date has passed) is one within the employer's control and which was caused by an event which would in any event have prevented the contractor from completing by the due date (provided of course that the employer was not relying on the contractor's progress in order to comply with a contractual, or statutory provision). Possible circumstances which give different results are given in Chapter 5.

If such delays cannot be prevented, careful monitoring and records are vital where there are several causes of delay after the completion date has passed.

4.11 Minimising exposure to claims: prevention

Stringent notice provisions and requirements to give particulars may be effective in avoiding claims by contractors who do not follow such provisions. However, this may increase the contract price and lead to conflict throughout the contract.

Whether, or not, there are sensible contractual provisions, and whether, or not, the contractor complies with them, the employer's professional advisors can minimise exposure to claims by ensuring that they do not cause delay by matters within their control (such as issuing late information). It is a mistake to assume that information can be delayed on the grounds that the contractor is in delay and is not ready for it. In many cases the contractor will be able to make out a case for an extension of time (or even time at large), particularly if the information is received at a time when it can be shown that it would have been impossible to complete the works by the due date having regard to all of preceding activities (see Figure 4.3). Scheduling issuance of information in accordance with the contractor's progress is a recipe for disaster and to be avoided at all costs.

Where delay and/or disruption claims occur, careful attention to records and constant monitoring of the effects will enable the employer to minimise his exposure. Inflated, or exaggerated, claims can be refuted. Costs which are partly to be borne by the contractor can be identified and adjustments made (see Chapter 7 – concurrent delays). Even where delays on the part of the employer justify an extension of time, the contractor's claim for payment can be reduced, or disallowed, where it can be shown that the contractor was also in delay and the costs claimed would, in any event, have been incurred by the contractor.

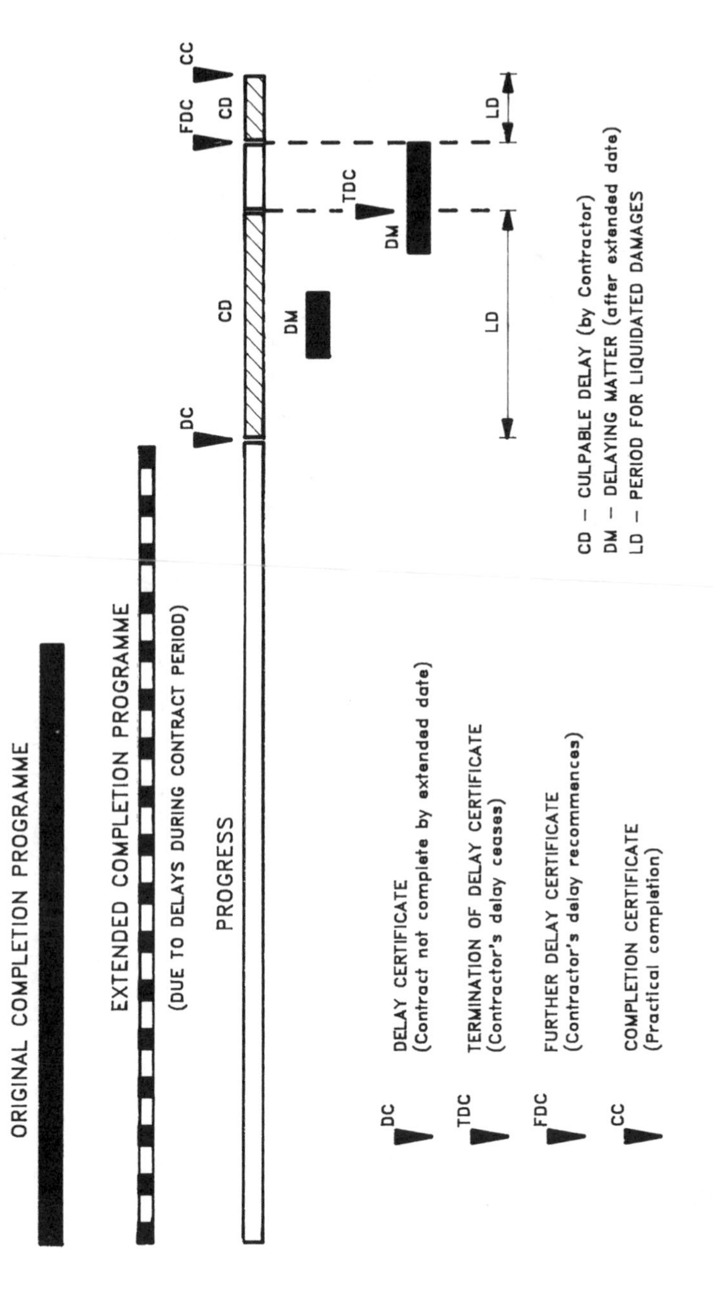

Figure 4.2 Clause 24 – Delay in completion and liquidated damages; SIA form of contract

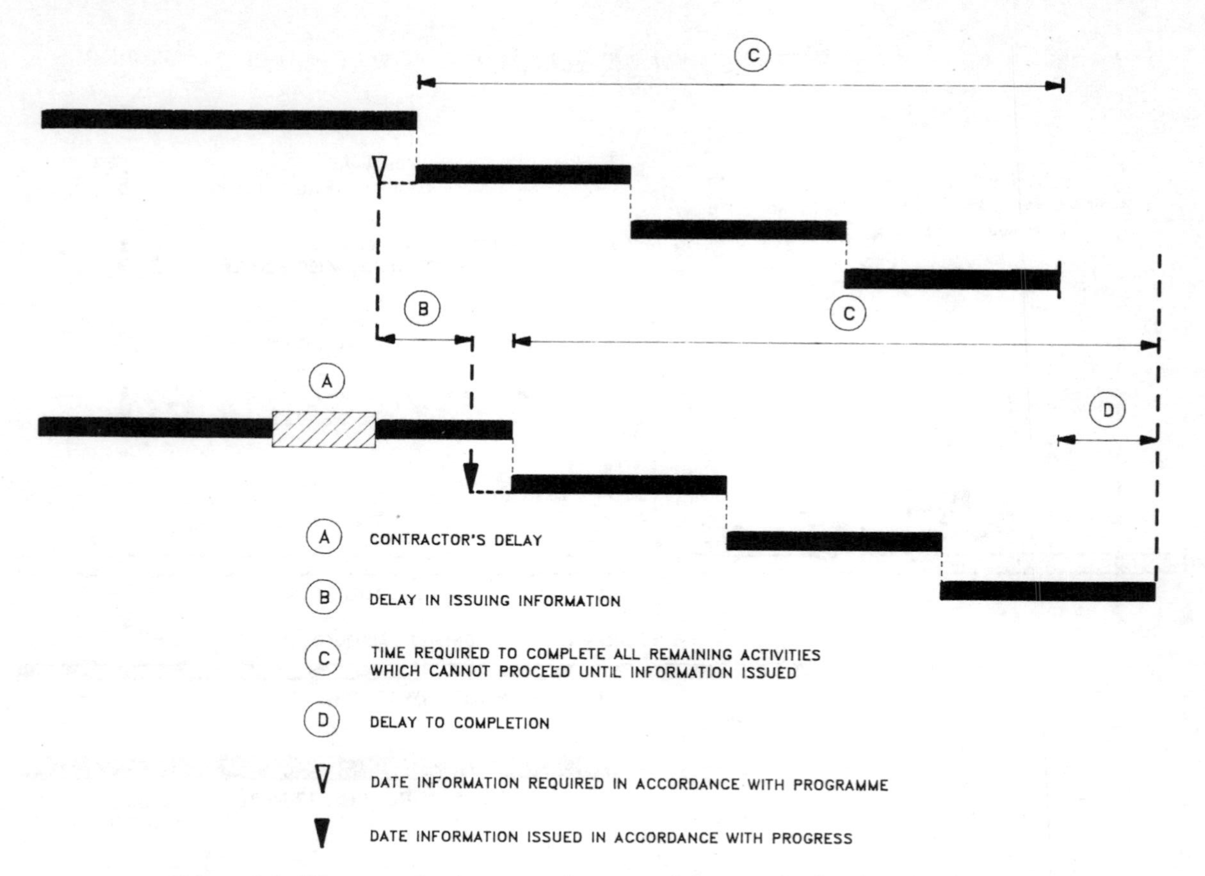

Figure 4.3 Time required to complete remaining work after late instruction

Delays and claims arising out of them are almost inevitable in construction contracts. If this fact is acknowledged, and proper procedures are devised to deal with them, then claims would be more palatable to those having to pay for them. Usually, all parties are at fault to a varying degree, and adversity thrives on one or more parties attempting to place all of the blame on someone else. Contractual provisions do not, in themselves, avoid these problems. Education and training in contracts administration should be encouraged to improve the understanding of claims and how they arise.

5 Formulation and Presentation of Claims

5.1 Extensions of time claims

All modern building and engineering contracts contain provisions for extensions of time in the event of delay. The nature of the work and the environment in which the work is carried out is such that it is almost inevitable that events and circumstances will cause completion of the work to be delayed beyond the original completion date. Notwithstanding, claims for extensions of time probably cause more disputes than any other contractual or technical issues. Major obstacles to prompt settlement of claims for extensions of time claims are:

- The erroneous assumption that an extension of time is automatically linked to additional payment;
- Late, insufficient or total lack of notice of delay on the part of the contractor;
- Failure to recognise delay at the appropriate time and maintain contemporary records;
- Failure to regularly update the programme so that the effects of delay can be monitored against a meaningful 'programme of the day';
- Poor presentation of the claim to show how the progress of the work has been delayed;
- Insistence, on the part of the employer's professional advisers, that unreasonably detailed critical path programmes are essential in order to assess the effects of the delay;
- The probability that the cause of the delay will reflect on the performance (or lack of it) on the part of the the employer's professional advisers;
- Pressure, on the part of the employer, to complete on time, irrespective of delays which occur.

The first obstacle – delay means money – is understandable. Nevertheless, it should not be a consideration when dealing with extensions of time. It should be clearly understood that an extension of time merely enables the contractor to have more time to complete the works and the employer to preserve his rights to liquidated damages. An extension of time awarded for a cause of delay which appears to have a financial implication (delay within the control of the employer) does not necessarily lead to an entitlement to additional payment. If the contractor is, himself, also in delay, then the additional costs arising out of the extended period to execute the works may (in total or in part) have to be borne by the contractor (see concurrent delays – *infra*).

On the other hand, an extension of time awarded for neutral events (for example adverse weather conditions) will not necessarily deprive the contractor of a claim for additional payment. The latter point was clearly illustrated in the case of *H. Fairweather & Co Ltd* v. *London Borough of Wandsworth (supra)*. In this case the arbitrator had concluded that the architect had been correct in awarding eighty-one weeks extension of time for the dominant cause of delay (strikes). The arbitrator had stated that the extension did not give rise to a claim for direct loss or expense. The contractor sought to establish that eighteen weeks extension of time ought to have been granted for causes of delay which would give rise to a claim for loss or expense.

The contract was JCT63 in which the some of the causes of delay (or disruption) in the loss and expense clause (24) are set out almost verbatim as some of the causes of delay in the extension of time clause (23). This is unfortunate and misleading and may be one of the reasons for some practitioners to assume a link between extensions of time and claims for additional payment. This misconception was cleared up by Judge Fox-Andrews QC in a hypothetical example which is summarised below:

> A tunnelling contract proceeds through the winter and is due to complete on 31 July. A variation instruction is issued in April which requires a further three months for completion of the works and for which an extension of time is granted up to 31 October. Two weeks before the revised completion date a strike occurs which continues until 31 March. The works cannot proceed and time passes through a second winter. On 1 April, the contractor recommences work, but due to the fact that it had not been able to protect its plant and equipment during the strike it takes two months to complete the remaining work. An extension of time for eight months for the strike (under clause 23(d) of JCT63) would not prevent the contractor from recovering loss and expense under clause 11(6). (See Figure 5.1.)

Nevertheless, in the circumstances of the case, the judge recognised the practical difficulties in the event of the extension of time not being made under the provision which linked the extension to the provisions of clauses 11(6) and/or 24(1) and he remitted the matter to the arbitrator for further consideration. It should be noted that clause 26.3 of JCT80 contains provisions which suggest a link between a claim for loss and/or expense and certain extensions of time made under clause 25. Whilst this may be

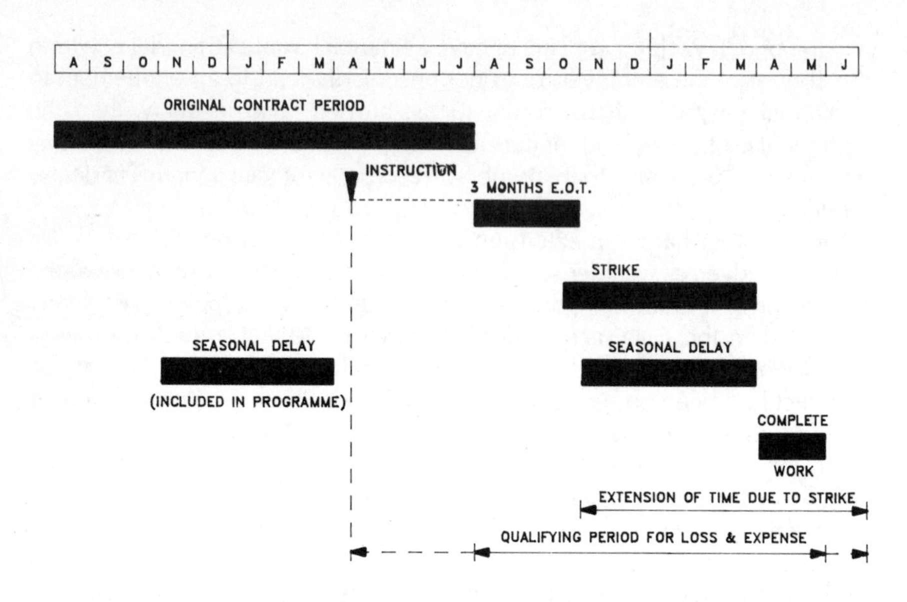

Figure 5.1 Fairweather & Co Ltd *v.* The London Borough of Wandsworth

desirable from a practical point of view, practitioners should not be misled into assuming that an extension of time for the specified relevant events will bring with it an entitlement to additional payment.

The next three obstacles, notice, contemporary records and programme, are all practical matters which can only be addressed by ensuring that adequate contracts administration procedures are being followed from the date of commencement of the works. Whilst the architect, or engineer, must do their best to estimate the length of any extension of time which may be due, irrespective of the lack of notice and particulars given by the contractor (*Merton* v. *Leach – supra*, Chapter 1), contractors cannot complain if the extension made on the basis of inadequate information does not live up to their expectations.

5.2 Presentation of extensions of time claims

Most contracts do not require the contractor to do more than give notice of delay, maintain records and provide particulars. Notice provisions vary. Some examples are:

- JCT80 – '...whenever it becomes reasonably apparent that the progress of the Works is being or is likely to be delayed the Contractor shall *forthwith* give written notice...'(Clause 25.2.1).
- GC/Works/1, Edition 3 – Notice may be given at any time, but not

'...after completion of the Works.' (Clause 36(4)). Clause 35 contemplates regular review of extensions of time, but there is no link to clause 36.

- ICE fifth edition – *Full and detailed particulars* '...shall be given within 28 days after the cause of the delay has arisen or as soon thereafter as is reasonable in all the circumstances...' (clause 44(1)). Similar provisions appear in the sixth edition.
- JCT80 goes on to require the contractor to give particulars of the expected effects of the delay (clause 25.2.2.1) and an estimate of the extent of any delay in completion of the works beyond the completion date (clause 25.2.2.2). GC/Works/1 requires the contractor to keep records (clause 25).

None of the above provisions requires the contractor to show the effects of the delay or to how it arrived at its estimate of the period of delay. Provided that the contractor has provided details of all events, dates, what work was affected and the like (together with an estimate of the delay in the case of JCT80), it appears that the contractual provisions have been satisfied and the onus is then on the architect, or engineer, to decide what extension is reasonable on the basis of the particulars provided and/or on the basis of further information obtained from other sources. Many contractors only provide information (often insufficient) and rely on the architect, or engineer, to make a reasonable extension of time. This tactic can be successful, but there is a risk that the extension made will be insufficient. Not all is lost, as the contractor can always present his case at a later date, hoping to persuade the opposition that more time is justified. The problems with this approach are:

- It is usually more difficult to persuade someone to change their mind after they have made a written extension of time unless there is additional evidence which can be used to explain a change in the period of the extension;
- There will almost certainly be a period of protracted discussion during which the current (extended or otherwise) completion date and the progress of the works are inconsistent with a realistic programme and a subsequently revised extended completion date.

These problems must be avoided or their effects will be compounded, making it difficult to monitor future delays and to make realistic extensions of time having regard to all of the circumstances. The better approach, on the part of the contractor, is to present his claim for an extension of time showing how he arrived at his estimate of delay and the effects on completion of the works. If the contractor has a detailed critical path programme using one of the well tried software packages, or a tailor-made package, then this task can be simplified. Unfortunately, many contractors who use such packages become complacent, believing that the programme, and the software used, is the answer to all of their problems. Computer applications can only be truly effective if the delays are quickly identified and steps are taken immediately to monitor events and update the programme. In many instances, full-blown computer applications are not necessary. Carefully prepared linked bar chart programmes can be very effective provided that the original logic is right.

Example 1 – A single cause of delay on the critical path

A linked bar chart showing how the contractor intended to complete the works in twenty-two weeks is shown in Figure 5.2.

A qualifying delay (D1) of two weeks occurred during weeks six and seven affecting progress of activity B-E (which is on the critical path – see Figure 5.3). In these circumstances it is a relatively simple matter to recognise that completion of the works was likely to be delayed by two weeks and an extension of time should be made for the full period of delay giving a revised completion period of twenty-four weeks.

The above example is straightforward as it deals with delay which is on the critical path and there are no concurrent delays. What is the situation in the event of delay which is not on the critical path? Some authorities exist which may be of some assistance.

In *Glenlion Construction Ltd* v. *The Guinness Trust (supra)*, the judge had to consider matters of extensions of time where the contractor had prepared a programme showing completion of the works before the contractual date for completion. Tenders were invited on the basis of a contract period of 104 weeks. *Glenlion* submitted an alternative tender for completion in 114 weeks which was accepted by *Guinness*. The completion date inserted in the contract was 114 weeks after the date for possession. The contract required *Glenlion* to produce a programme showing completion 'no later than the date for completion' and *Glenlion* complied by producing a programme which showed completion in 101 weeks. There were delays and disputes arose as to *Glenlion's* entitlement to an extension of time. The crucial text of the judgement is (at page 104):

> 'Condition 23 [extensions of time] operates, if at all, in relation to the date for completion in the appendix. A fair and reasonable extension of time for completion of the works beyond the date for completion stated in the appendix *might be* an unfair and unreasonable extension from an earlier date.' (Emphasis added).

It must be concluded that if any delay occurs then it is not necessarily correct to make an extension of time equal to the period of delay. Some, or no, extension of time may be required. How much extension (if any)? The following quote from *Hudson's Building & Engineering Contracts, Tenth Edition, First Supplement* at page 639 may be helpful:

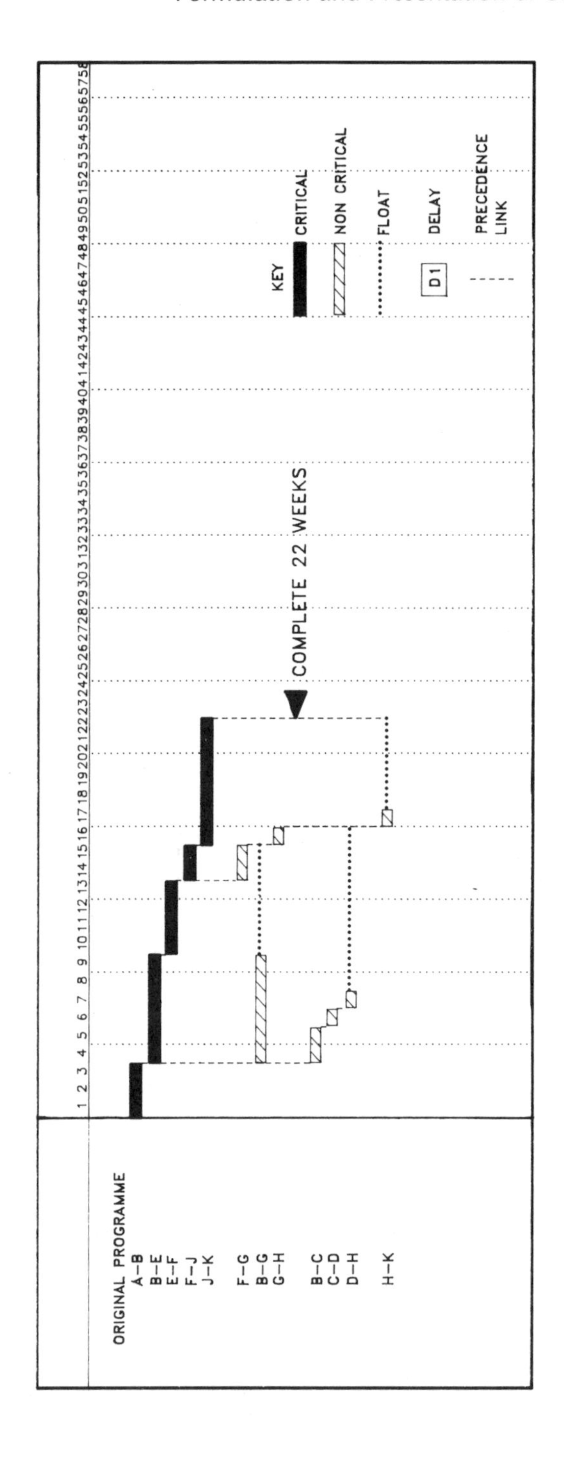

Figure 5.2 Precedence (linked) bar chart – original programme

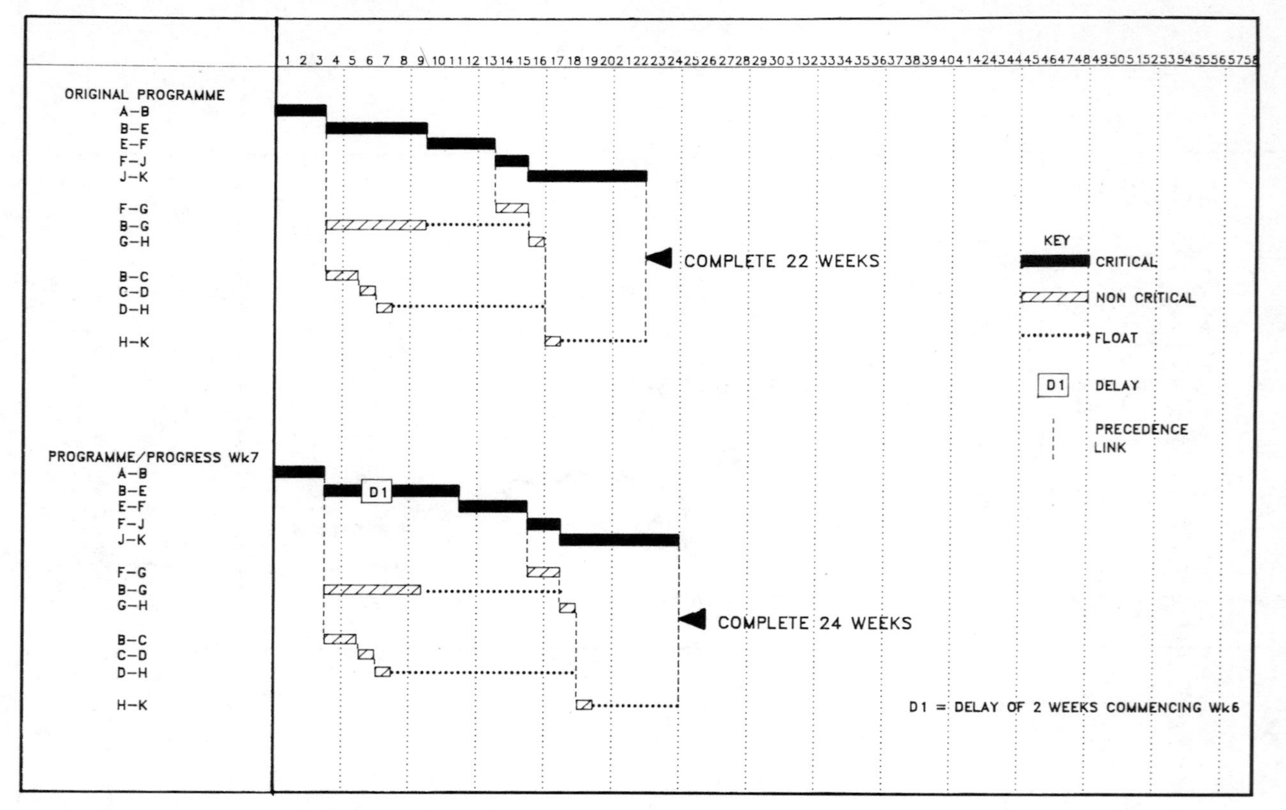

Figure 5.3 Single cause of delay on the critical path

> '... a contractor may be *in advance* of planned progress and an event justifying an extension will only have the effect of his losing that advantage, should some later default occur, but not imperil the actual date. Ideally such an extension need only be given if the contractor later has need of it – i.e. by being in culpable delay...'.

The above quote from *Hudson* confirms the widely held view that any float in the contractor's programme is for the benefit of the contractor and any delay on the part of the employer which reduces that float may have to be taken into consideration when considering the time required for completion.

This concept can be applied to *Glenlion* v. *Guinness* as shown in Figure 5.4. Bar A indicates the period for completion stated in the tender documents (104 weeks), bar B indicates the period for completion stated by *Glenlion* in the alternative tender (114 weeks, which was accepted by *Guinness*) and bar C indicates the period indicated in *Glenlion's* programme (101 weeks). The programme shows completion thirteen weeks before the contractual date for completion.

Assume that a delay of five weeks occurs at the outset of the contract for which there is power to make an extension of time (that is, a qualifying delay or relevant event – bar D). This has the effect of reducing the contractor's float from thirteen weeks to eight weeks. No extension of time is necessary as completion is not likely to be delayed beyond the contractual date for completion.

A further qualifying delay of four weeks occurs during the contract period (bar E). Again, this only reduces the contractor's float from eight weeks to four weeks and no extension of time is necessary. Another qualifying delay of four weeks occurs towards the end of the contract which takes up the remaining float (bar F). Again, no extension of time is necessary.

Four weeks before completion, a further delay of four weeks occurs which does not qualify for an extension of time (for example culpable delay on the part of the contractor). In these circumstances the contractor has need of an extension of time and it would therefore be reasonable to make an extension of time of four weeks. Difficulties may arise under JCT80 because the extension of time clause (25.3.1) contemplates an extension of time being made if '...completion of the Works is likely to be delayed [by the relevant event] beyond the Completion Date...'. In the above example, completion of the works was delayed beyond the completion date by an event which did not qualify for an extension. However, the circumstances described in this example may be covered by the provisions of clause 25.3.3 which empowers the architect to '...fix a Completion Date later than that previously fixed if in his opinion the fixing of such later Completion Date is fair and reasonable *having regard to any of the Relevant Events...*' (emphasis added). Some may argue that clause 25.3.3 does not apply in these circumstances. Even if that view were to be correct, the employer would be unlikely to succeed in

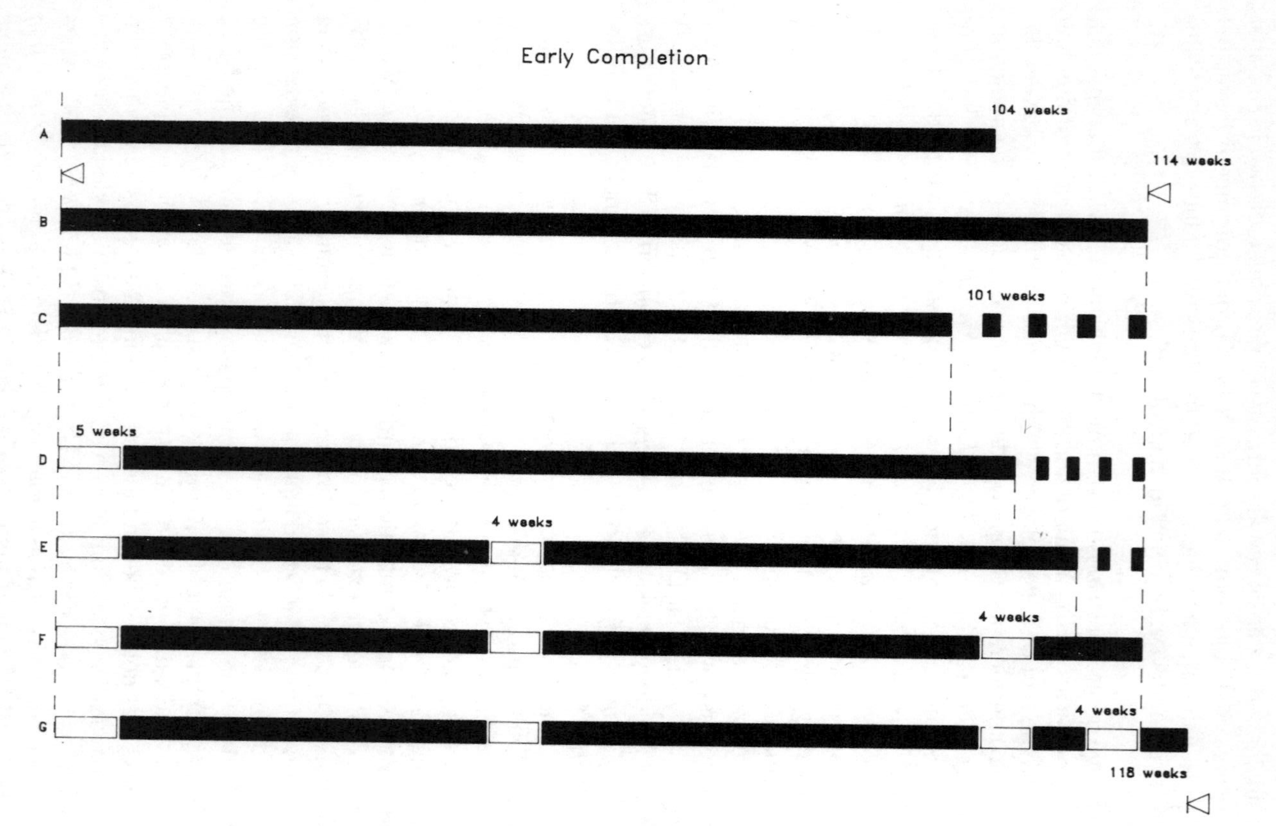

Figure 5.4 Glenlion Construction Ltd *v.* The Guinness Trust

claiming liquidated damages for late completion when it has been partly responsible for the delay to the progress of the works. Regard may have to be paid to the nature of the contractor's culpable delay. Sheer dilatoriness on the part of the contractor may be viewed in a different light to matters such as a plant breakdown or failure to obtain materials in spite of taking all reasonable measures.

Those who resist making an extension of time in circumstances similar to the above example may be persuaded to change their view by considering the position if any (or all) of the delays in bars D, E and F had been due to the contractor's own delay and the delay in bar G had been due to a qualifying delay. In these circumstances, there is no room to argue that an extension of time is not required. This would appear to be the case even if the contractor's own delays had been due to dilatoriness, since the contractor would not be in breach of its obligation to complete until the completion date had passed.

Note – Clause 33 of GC/Works/1, Edition 3, requires the contractor's programme to '...use the whole period for completion.'.

Example 2 – A single cause of delay – not on the critical path

Using the same linked bar chart in Figure 5.2, a qualifying delay (D2) of two weeks occurred during weeks six and seven which affected the progress of activity B-G (which is not on the critical path – see Figure 5.5). In these circumstances there is no effect to the completion date and no extension of time is necessary.

Example 3 – Concurrent delays – critical and non-critical

Using the same linked bar chart in Figure 5.2, the delays referred to in examples 1 and 2 above occurred at the same time (see Figure 5.6). If both of the delays were qualifying delays, an extension of time of two weeks is necessary for the delay (D1) which affected activity B-E. If the delay to activity B-E is a qualifying delay, and the delay (D2) to activity B-G is due to the contractor's culpable delay, an extension of time of two weeks is necessary. This is the case even when it is clear that the concurrent delays are operating during identical periods. This would also be the case if the contractor's culpable delay (D2) to activity B-G was on a parallel critical path and therefore also delaying completion by two weeks.

If the delay (D1) to activity B-E was due to the contractor's culpable delay, and the delay (D2) to activity B-G was a qualifying delay, then no extension of time would be necessary.

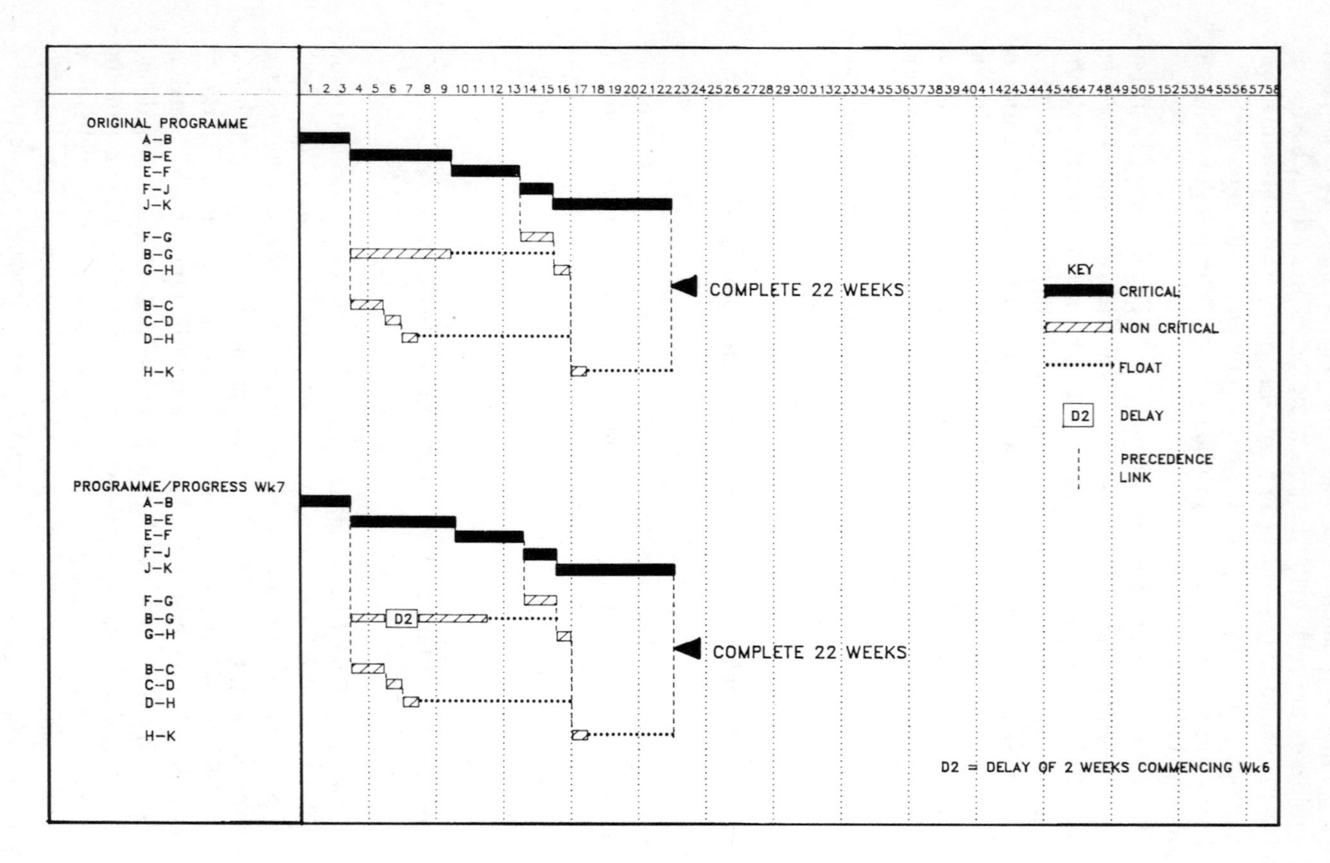

Figure 5.5 Single cause of delay – non-critical

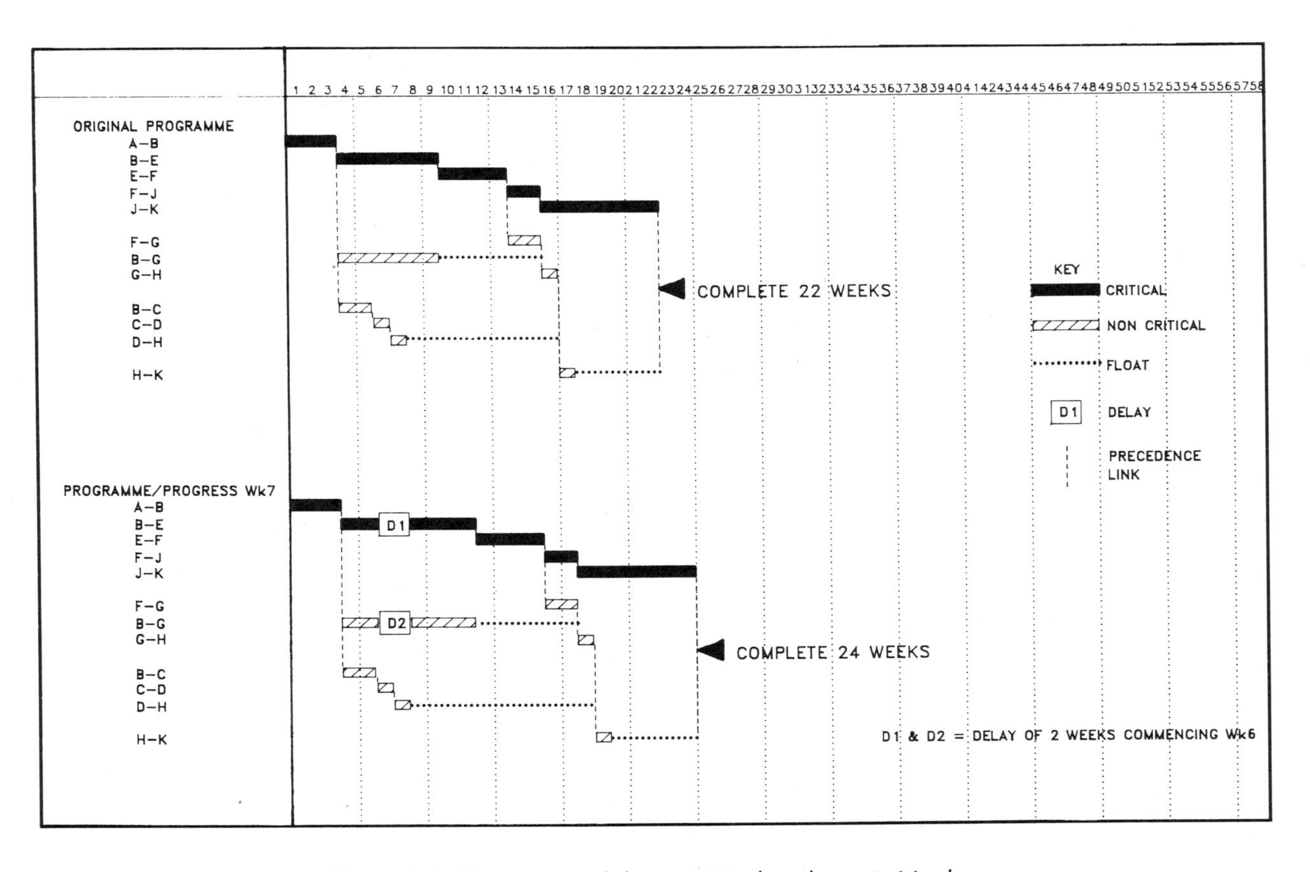

Figure 5.6 Concurrent delay – critical and non-critical

Example 4 – Concurrent delays followed by subsequent delays

Using the same linked bar chart in Figure 5.2, the delays referred to in examples 1–3 above were followed by further delays of seven weeks (D3) and five weeks (D4) to activities B-G and H-K respectively. If delays (D1) and (D2) were both qualifying delays (or if delay D2 was a non-qualifying delay), an extension of time of two weeks should already have been made (completion in twenty-four weeks). If delay (D3) was also a qualifying delay it would have the effect of delaying commencement of activities G-H and H-K, but no extension of time would be necessary because the float allowed for activity H-K is more than sufficient to absorb the delay (the float is reduced from five weeks to four weeks – see Figure 5.7).

However, for the reasons given previously, if delay (D4) occurred due to some event which did not qualify for an extension of time (for example, non-availability of materials, such as road surfacing, which could not be stored on site for use) an extension of time may be necessary *because the contractor had need of it* (see Figure 5.8). In these circumstances, qualifying delays (D2) and (D3) had reduced the contractor's float and non-qualifying delay (D4) had used up more than the remaining float, thereby causing completion to be delayed by one week (completion in 25 weeks). If delays (D2) and (D3) had not occurred, there would have been sufficient float remaining in activity H-K to absorb the delay (D4) and there would have been no delay to completion beyond the previously extended completion period of twenty-four weeks.

Numerous permutations may arise and each delay and its effects on the remaining float and the completion date need to be considered using the principles described above.

5.3 Delays after the contract completion date

It is well known that the extension of time provisions of JCT63 (clause 23) do not deal with delays which occur after the contract completion date (extended or otherwise) has passed and the contractor is in culpable delay. Indeed the clause is drafted in terms which appear to preclude making an extension of time for any delay which occurs after '...any extended time [date] *previously fixed...*' (emphasis added). That is to say, even if an extension of time ought to have been made for previous delays, if the extension has not been made by the (then) current extended completion date, and a new (otherwise qualifying) delay occurs, there is no power to extend time for completion. This situation does not appear to be capable of rectification by subsequently making an extension of time for the previous delay, thereby causing the new delay to occur before the subsequently revised extended completion date.

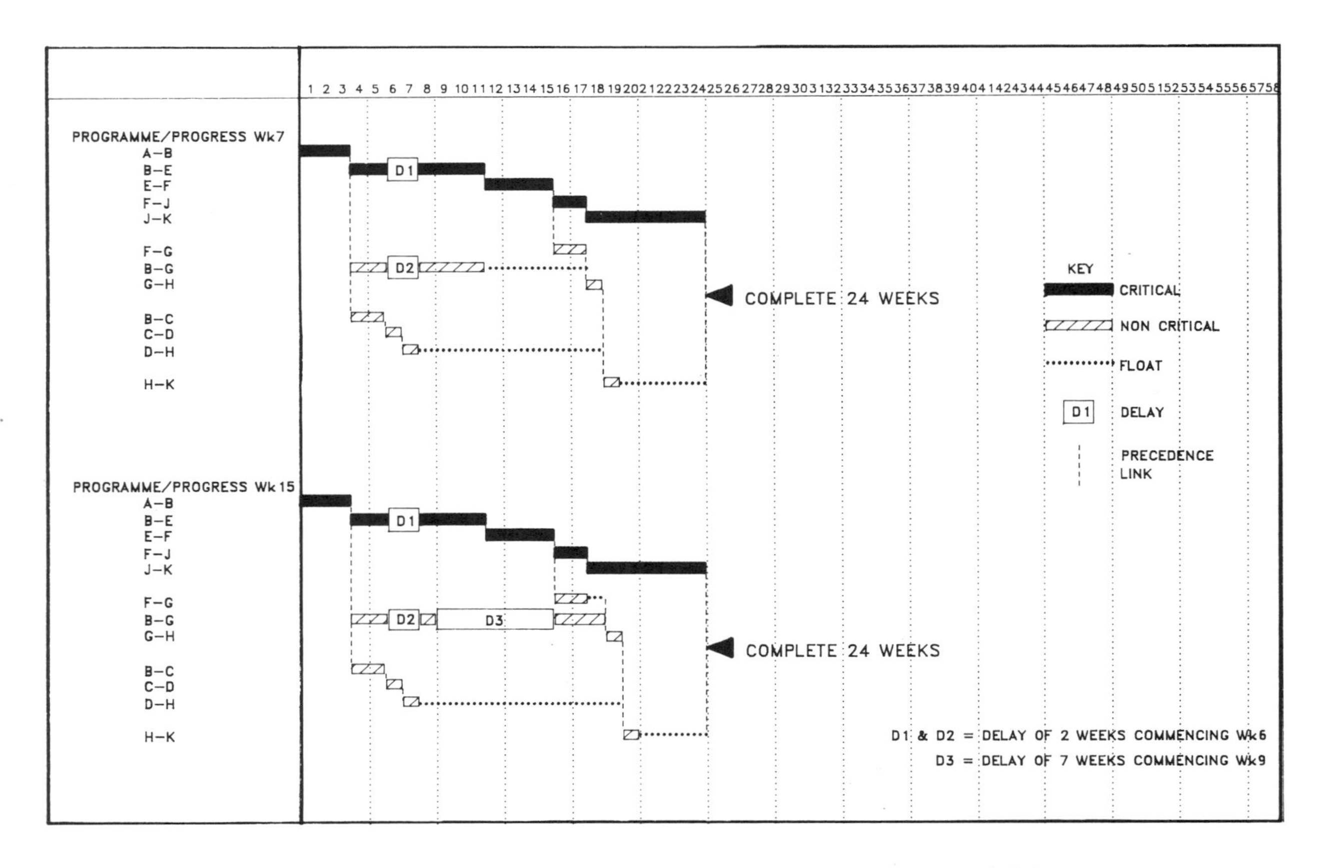

Figure 5.7 Concurrent delay followed by subsequent non-critical delay

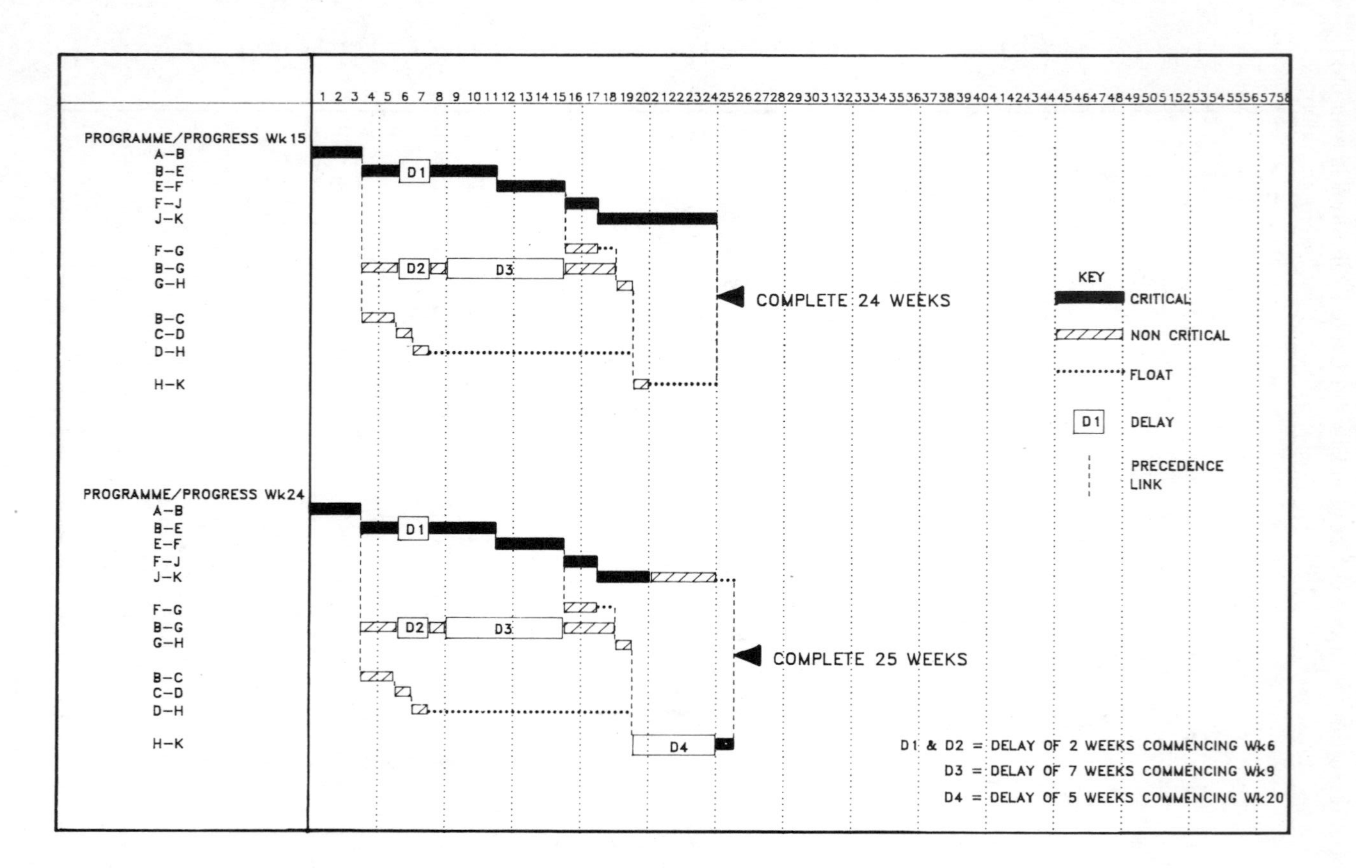

Figure 5.8 Concurrent delays followed by subsequent non-critical and critical delays

It is doubtful if any current contract in the United Kingdom is executed under the terms of JCT63. However, extensions of time provisions identical to JCT63 are still in everyday use in many parts of the world. Bahrain, Cyprus, Hong Kong and Jamaica are a few examples. Wherever these contracts are in use, it is therefore essential to make extensions of time for all known delays (whether, or not, notified by the contractor) before the existing completion date has passed. Failure to do so may cause time to be at large and invalidate the liquidated damages provisions.

Problems associated with delays after the completion date are not confined to JCT63, *Hudson's Building & Engineering Contracts, Tenth Edition, First Supplement* at page 653:

> 'One further matter not covered by the vast majority of extension of time clauses is whether they are intended to operate during a period of culpable delay in respect of matters which, but for the contractor being in delay and already liable for liquidated damages, would entitle the contractor to an extension. Careful analysis shows that, if so, additional machinery is required... No UK standard form as yet contains any such provision.'

The distinguished author of *Hudson* has gone to great lengths to introduce the necessary 'additional machinery' in clause 24 of the form of contract issued by the Singapore Institute of Architects. It is not considered to be necessary to deal with this clause at length in this chapter. However, a diagram showing how the clause is intended to operate is shown in Chapter 4 (see Figure 4.2 – *supra*).

Other widely used forms of contract at the time of publication of the First Supplement to *Hudson* were, the fifth edition ICE conditions of contract, third edition FIDIC, GC/Works/1 Edition 2 and a few minor works forms of contract. These forms of contract do not appear to prohibit extensions of time after the completion date has passed. However, the provisions are unclear and there is no guidance as to the period of extension, and its effect on the employer's rights to liquidated damages. Later forms of contract, such as JCT80 and fourth edition FIDIC, offer nothing to assist in this situation. The Intermediate Form of Contract (IFC84) expressly provides for extensions of time to be made for delays which occur after the completion date has passed, but there are no rules setting out how this should be done.

These problems are addressed in the following example (see Figure 5.9).

In this example it can be seen that a delay (D1) which occurs before the contract completion date is capable of being dealt with by an appropriate extension of time. A new completion date (NCD1) can be fixed according to the circumstances.

When a new qualifying delay (D2) occurs after the completion date has passed and the contractor is in culpable delay, what period of delay should qualify for an extension of time? Should it be the total period of delay (TD)

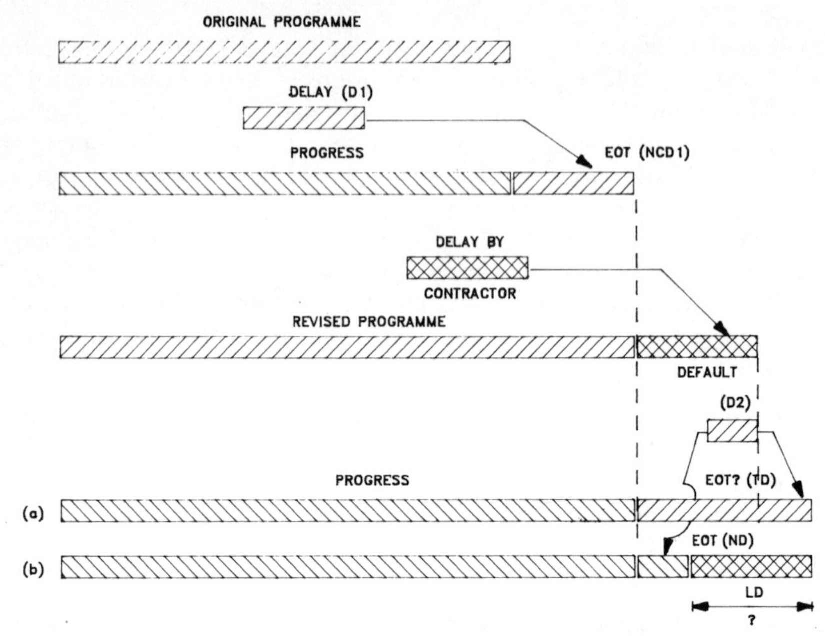

Figure 5.9 Delay by employer after completion date

from NCD1 to the earliest completion date caused by the new qualifying delay or should it be for the nett period of the new qualifying delay (ND)? Can liquidated damages be levied?

Consider two possible alternatives:

Alternative A

Eight weeks after the contract completion date, the contractor commences excavation for the final connections to the foul drainage. The work ought to have been carried out not later than two weeks before the completion date. With the exception of delay (D1), there have been no delays for any reason other than the contractor's failure to proceed in accordance with its programme. Unknown existing gas main and power cables are discovered which necessitate a variation to change the routing of the drainage and the construction of an additional inspection chamber. The additional work causes a delay of one week (D2) and completion of the works is delayed by one week.

In these circumstances, had the contractor not been in culpable delay, the necessity for a variation would have come to light before the completion date and an extension could have been made at the time. Therefore, if the contractor had been proceeding in accordance with his programme, one week extension of time (beyond the date already fixed as a result of delay D1 – NCD1) would have been reasonable (ND).

Alternative B

In the same circumstances as alternative A, eight weeks after the completion date has passed, the contractor is instructed by the architect to cease work on the excavation for the foul drainage. The architect then instructs the contractor to vary the levels and diameter of the pipes and construct an additional inspection chamber and two additional branch connections for a future extension. The additional work causes a delay of one week (D2) and completion of the works is delayed by one week.

In these circumstances, the architect could, and ought to have, ordered the additional work in sufficient time to enable the work to be carried out before the completion date and without causing delay. The variations ordered by the architect were not dependent upon the contractor's progress and could not be attributable to the contractor's culpable delay. If the contract permitted an extension of time for delays which occurred after the completion date had passed, an extension of time for a period of ten weeks may be reasonable in the circumstances (TD).

Note – Most forms of contract (even if they contemplate extensions of time for qualifying delays occurring after the completion date has passed) do not contain the essential machinery to enable the employer to deduct liquidated damages for the period when the contractor is in culpable delay. As soon as the employer causes delay, the contractor's liability for liquidated damages may evaporate, leaving the employer to prove unliquidated damages (see Chapter 7).

5.4 Summary on presentation of extensions of time claims

In any claim for an extension of time, and whether or not there is a requirement to give details and particulars, it is good practice to include the following:

- A description of the cause of delay and the contractual provision which is being relied upon for the extension;
- The date when the delay commenced and the period of delay (giving details of intermittent effects if appropriate);
- The date of notice of delay, specifying the reference of the relevant document;
- A summary of records and particulars relied upon (with copies included in an appendix);
- A narrative of the events and effects on progress;
- A diagrammatic illustration showing the status of the programme, progress and current completion date prior to the commencement of the delay;

- A diagrammatic illustration showing the effects of the delay on progress and the completion date (including subsequent delays which may have reduced the float in the programme);
- A statement requesting an extension of time for the delay to completion for the period shown on the submitted illustrations.

5.5 Recovery of loss and/or expense and/or damages

Whilst failure to give notice of delay for extensions of time is not usually fatal to a claim, failure to give notice in accordance with the contract with respect to additional payment may bar, or severely prejudice a claim.

There are good reasons for contracts to have provisions for the contractor to give notice. No employer will wish to have a substantial claim appearing 'out of the blue' at the end of a contract. In *J. and J.C. Abrahams* v. *Ancliffe* [1938] 2 NZLR 420, a contractor estimated the cost of building two residential units at $30 000. Several months later the employer's architect issued a specification for the work and the contractor commenced work. It became evident that the specification provided for more expensive work than that which had been allowed for in the contractor's estimate. There were also problems in the foundations which increased the amount of work done and general building costs were escalating. The employer repeatedly asked the contractor for details of the expected costs but at no time did the contractor reply. When it came to settle the account the employer argued that the contractor was in breach of a duty to give reliable information about the costs of building before the employer became committed to completing the units at an uneconomic cost. It was held that the contractor was under a duty of care to the employer in giving its original estimate and to inform the employer as soon as it was aware that costs were going to substantially exceed the estimate.

In most forms of contract, the onus is not entirely upon the contractor to keep the employer informed of increases in the contract price. In most instances, the employer relies to a great extent on his professional advisers. In varying degrees (according to the terms of the contract) there must be co-operation between the employer's professional advisers and the contractor so that any increase in the contract price can be ascertained at the earliest possible time: *London Borough of Merton* v. *Stanley Hugh Leach Ltd* (*supra*). Where there are no express terms, co-operation is usually implied. Most construction contracts have express provisions making it clear as to what form this co-operation should take.

5.6 Notice of intention to claim

Most contractors do give notice of their intention to claim at some time during the contract. Some avoid any indication at all of their intention to claim until after an extension of time has been made. The former may barely comply with the contract and may prejudice the contractors' entitlements to some extent. The latter will invariably be the beginning of an uphill struggle to obtain payment of substantially less (if anything at all) than might otherwise have been possible if the contractor had given prompt notice. Notice provisions in modern construction contracts vary considerably:

- JCT80 – Clause 26.1.1 merely requires the contractor to make an application '...as soon as it has become, or should reasonably have become, apparent to him that the regular progress of the Works or of any part thereof has been or was likely to be affected [by the matters referred to]...'. It may be difficult to decide whether or not an application is late in all the circumstances. The only significant difference between the present clause and its predecessor (JCT63) is the addition of the words '...or should reasonably have become [apparent]...'. The clause lacks express language to bar a claim if an application is made 'late'.
- GC/Works/1, Edition 3 – Clause 46 (3) states that the contract sum *shall not be increased unless* '(a) the Contractor, immediately upon becoming aware that the regular progress of the Works or any part of them has been or is likely to be disrupted or prolonged has given notice to the [Project Manager] specifying the circumstances causing or expected to cause that disruption or prolongation and stating that he is,or expects to be, entitled to an increase in the Contract Sum...'.
- ICE fifth edition – Clause 52(4) requires the contractor to '...give notice in writing of his intention [to claim] to the Engineer as soon as reasonably possible after the happening of the events giving rise to the claim.' The sixth edition introduces a twenty-eight day period after the event giving rise to the claim has arisen, but like the fifth edition, if the contractor fails to comply with the contractual provisions, the contractor is entitled to payment so far as the engineer has not been prevented from investigating the claim.

5.7 Particulars and further information to support a claim

If proper notice has been given pursuant to the terms of the contract, both parties are aware of the claim and further steps can be taken to deal with it. Various provisions include:

- JCT80 – *If requested by the architect,* the contractor is required to submit appropriate information for the purposes of enabling the architect to form an opinion as to whether or not the contractor *has incurred or is likely to incur* direct loss and/or expense (clause 26.1.2) and *if requested by the architect or quantity surveyor,* the contractor is required to provide details of the loss and/or expense (clause 26.1.3). No time limits are specified for the architect's or quantity surveyor's requests or for the contractor's response.

- GC/Works/1 Edition 3 – The contract sum *shall not be increased unless* '(b) the Contractor, as soon as reasonably practicable, *and in any case within 56 days of incurring the expense,* provides full details of all expenses incurred and evidence that the expenses directly result from the occurrence of one of the events...' (clause 46(1)).
- ICE fifth and sixth editions – Require the contractor to give a first interim account and details as soon as possible after giving notice, and thereafter further accounts at such intervals as the engineer may reasonably require (clause 52(4)).

It appears that, with the exception of GC/Works/1, there is no bar to a claim provided that notice and particulars are given within a reasonable time.

Notwithstanding the loose provisions which appear to prevail, contractors are advised to give prompt notice followed by detailed particulars backed up by adequate contemporary records.

The methods of illustrating delay and disruption in support of claims for additional payment are similar to those used for illustrating claims for extensions of time.

5.8 Prolongation claims

Qualifying delays on the critical path will usually support a claim for prolongation costs for the period of delay (if such delays are matters which give rise to additional payment). For the purposes of claims for additional payment the term 'qualifying delay' means delay which brings with it the right to additional payment (some qualifying delays for extensions of time, such as adverse weather conditions, do not normally give rise to additional payment). Typical heads of claim arising out of prolongation of the contract period are:

Site overheads or preliminaries

It is surprising how many claims are submitted on the basis that the extra site overhead costs due to prolongation are those incurred after the original contract completion date and up to the extended (or actual) completion date. This is, of course, incorrect, but it may explain why some contractors wait until the end of the project to give notice and submit a claim. The following example illustrates how prolongation costs may be significantly understated using the above assumption.

The qualifying delay on the critical path (D1) shown in Example 1 (see Figure 5.3) has caused the completion date to be delayed by two weeks. The actual weekly costs of the contractor's general site establishment (time related costs) are shown in Figure 5.10.

It will be seen that the weekly costs incurred during the two week period of overrun (CO) are much lower than the weekly costs during the period of delay (CD). It is the cost incurred during the period of delay which should be the basis of the contractor's claim for prolongation costs. A claim based on the costs incurred during the period of overrun will normally be substantially less than the actual costs incurred during the period of the delay.

The costs incurred during the period of delay may not reflect the true *additional* costs of the delay. For example, the contractor may have recruited an electrical engineer to commence on site in the ninth week to supervise the electrical installation. There may be no other site at which the engineer can be usefully employed and it may not be possible to postpone his employment. The delay may have caused the commencement of the electrical installation to be delayed by two weeks, in which case the contractor is faced with paying the salary of the engineer for two weeks (weeks nine and ten) when there is no work being done which requires the engineer's supervision. This additional cost is a direct result of the qualifying delay and ought to be recoverable. However, the cost of the engineer is not included in the costs incurred in weeks six and seven (the period of delay). In order to overcome such problems, the contractor should show the periods when every time related resource was on site (and their costs) and when they ought to have been on site (save for the delay) – see Figure 5.11.

In practice, some qualifying delays may occur in isolation (as in the previous example) and/or numerous qualifying delays may occur over a period in which each qualifying delay overlaps with other qualifying delays. The nett result of all of the qualifying delays may cause prolongation of the contract period. Providing that there are no major concurrent delays by the contractor (which would be a matter of evidence) it may be reasonable to base a claim for prolongation costs on the costs shown in Figure 5.12.

In the above example, the cost of the isolated delay (A) may be established using similar principles as the previous example. The costs arising out of the numerous continuing delays during the period (B) may be taken as four-tenths of the total costs incurred during period (B). Some adjustments may have to be made for special circumstances such as the case of the electrical engineer used in the previous example. Alternatively, comparison between the resources which were utilised on site and the resources which ought to have been utilised (save for the delay) may give a more accurate result.

In any event, it is not the comparison between the actual resources and

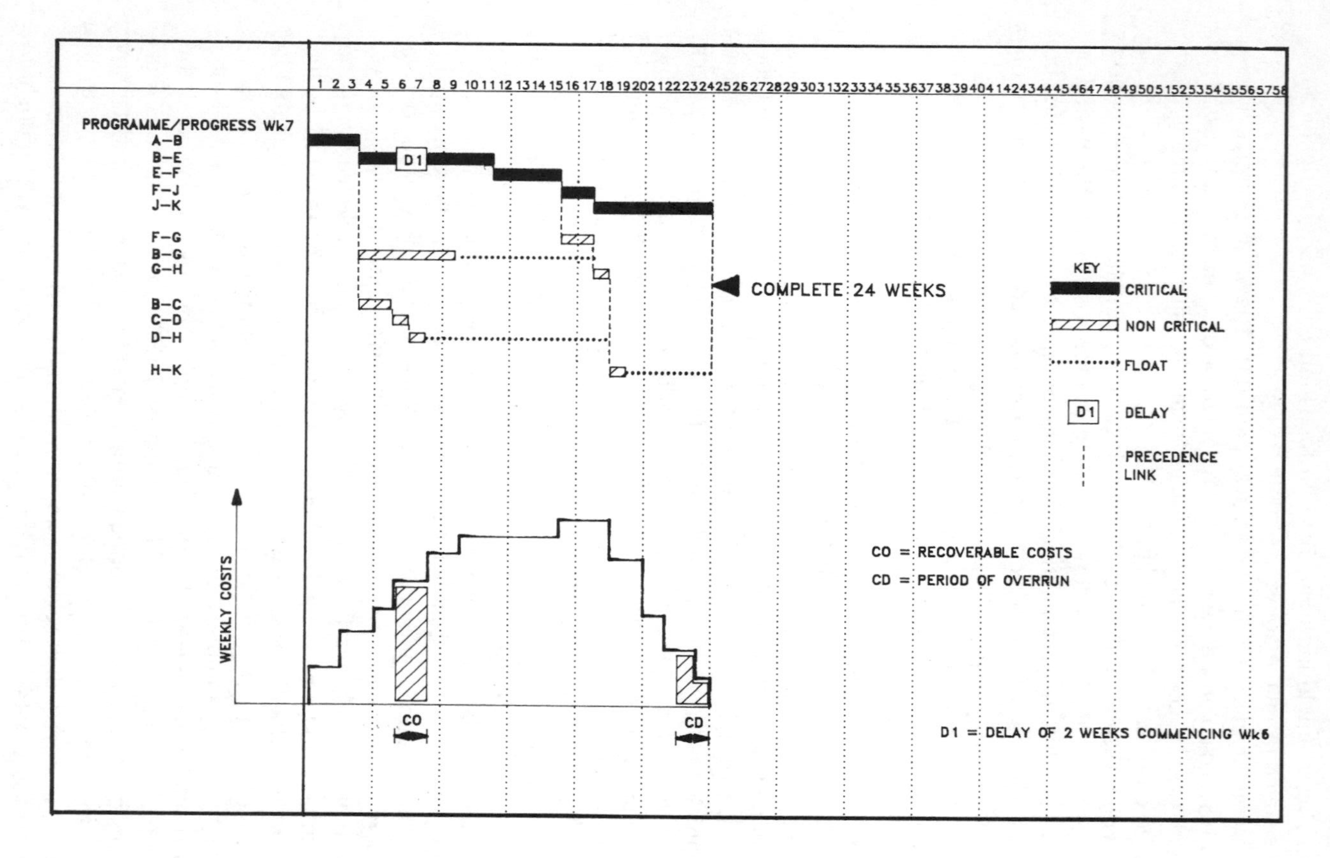

Figure 5.10 Recoverable site overheads

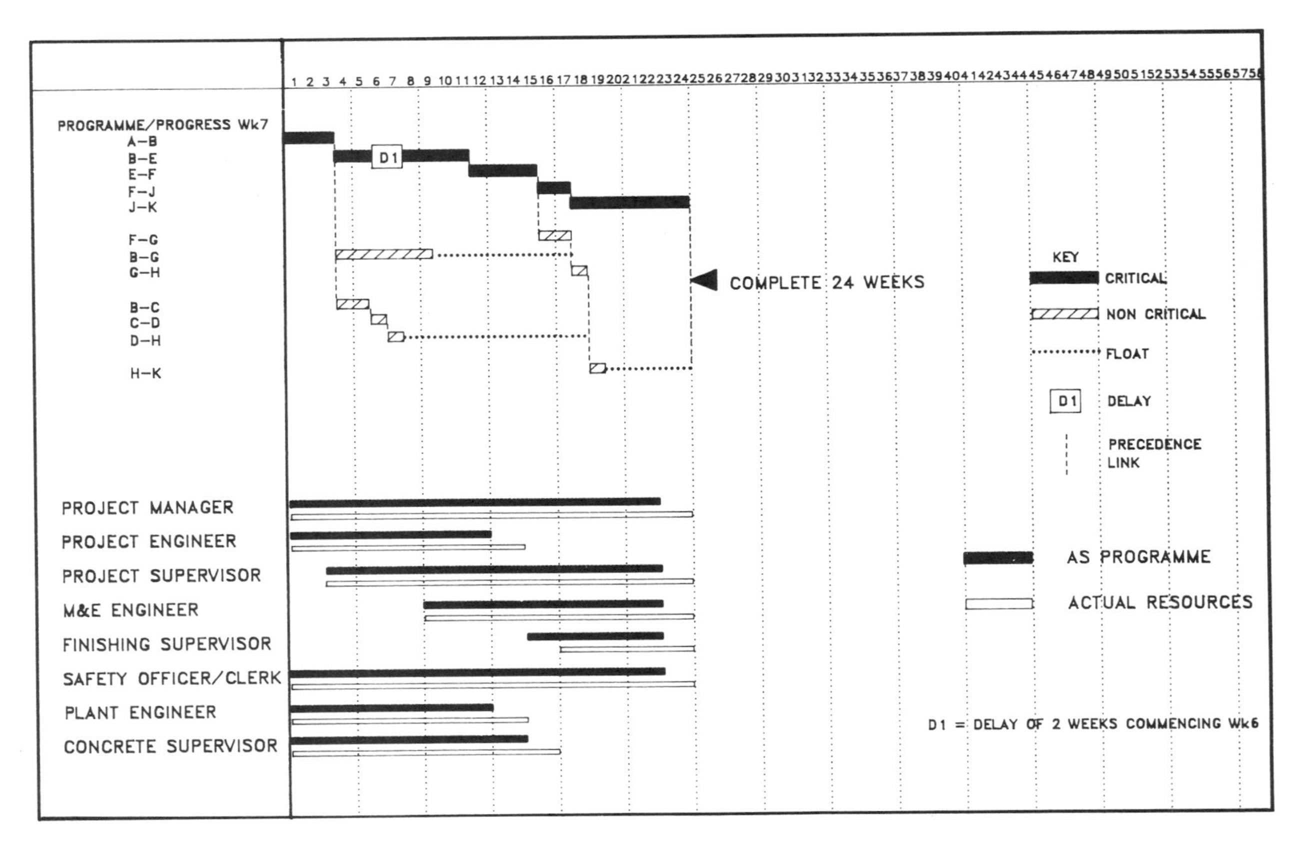

Figure 5.11 Anticipated and actual resources

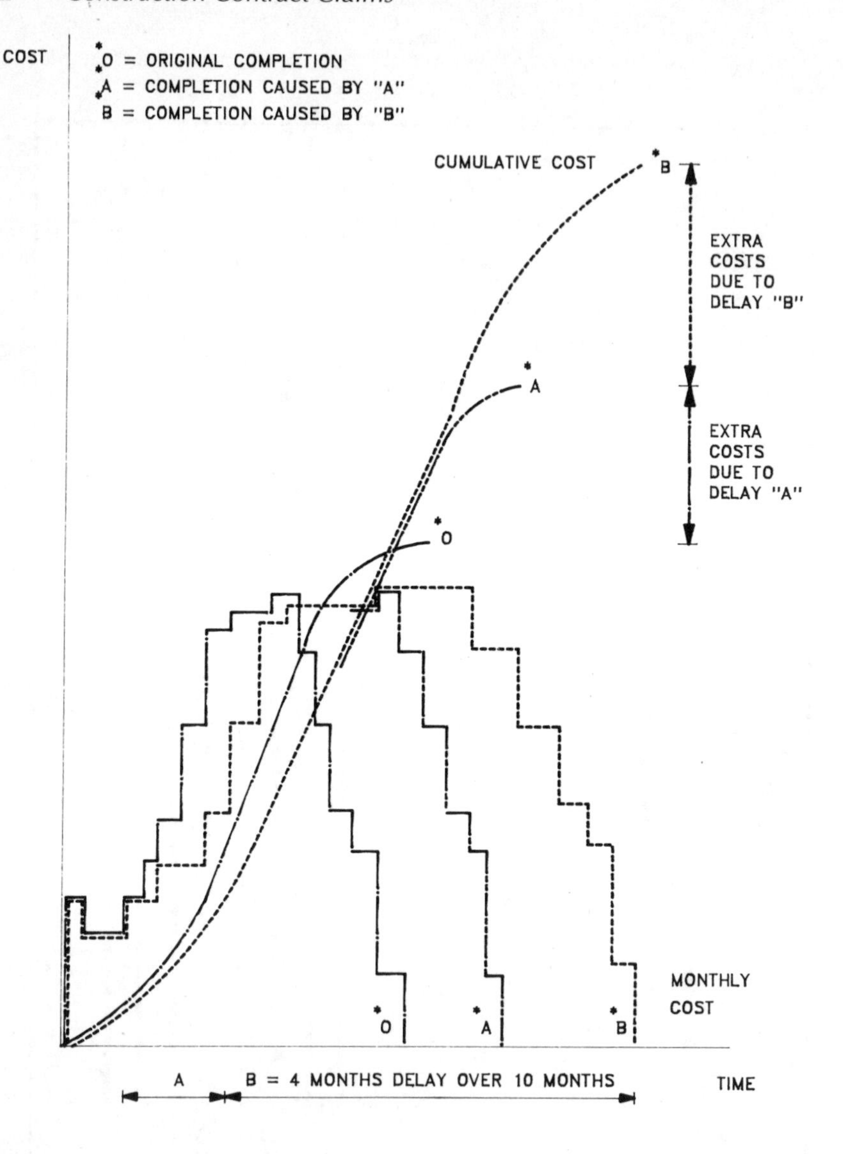

Figure 5.12 Extended preliminaries

those included in the contractor's tender which form the basis of the claim. If the contractor can show that it was reasonable and necessary to employ more weekly resources than those allowed in the tender he may be able to claim on the basis of the increased resources. However, if there was no good reason to employ additional resources, the contractor's claim may be limited to the costs of resources which were consistent with the contractor's tender assumptions. If the contractor's actual resources were less than the tender provisions, the employer would not expect to reimburse the contractor any more than the actual costs incurred.

Prolongation of individual activities

Some delays may not be on the critical path, in which case there will be no general prolongation costs. However, some time related costs may be solely attributable to a particular activity. If delay (D2) in example 2 (see Figure 5.5) is in respect of an activity which requires scaffolding for its total duration, then the cost of the scaffolding for the period of the qualifying delay of two weeks would be recoverable. Supervision and other plant and equipment utilised solely for the activity may also be recoverable. This is particularly valid where the activity is for work carried out by a subcontractor. The subcontractor will have a prolongation claim against the contractor and the contractor will seek reimbursement under the relevant provisions of the principal contract.

Valuation at cost or using contract rates for preliminaries

If the delay was caused solely by a variation, it could be argued that the valuation of the variation should take into account the time related rates in the contract bills (see Variations – *infra*). Account would have to be taken of significant changes in actual costs when compared with the time related rates in the contract bills. If the delay was caused by breaches of contract, such as late issuance of drawings and details, the remedy is by way of damages, thereby requiring the loss to be based on the contractor's actual costs irrespective of the contract rates. If the delay was caused by variations and breaches of contract, and the periods of delay for each cause cannot be disentangled, it is suggested that actual costs should be used as the basis of any claim.

Head office overheads in the event of prolongation

Various formulae may be used. However, some doubt was cast upon the use of a formula in *Tate & Lyle Food Distribution Ltd and Another* v. *Greater London Council* [1982] 1 WLR 149. It should be noted that in this case very little evidence (if any) was put forward to establish the extent of disruption and delay and there was no evidence presented to support the percentage claimed. It is thought that where a contractor can show evidence of delay, and the extent of it, and where there is evidence to support the contention that resources were prevented from earning a contribution to overheads and the percentage to be used, then one of the recognised formulae may be used.

The Hudson formula

This formula was put forward in *Hudson's Building and Engineering Contracts, tenth edition* 1970 (page 599). It uses the percentage in the contractor's tender for overheads (and profit, if applicable) as a basis for the contractor's loss of contribution to overheads (profit), as a result of delay, in the following formula:

$$\frac{\text{H.O. Overheads (profit)\%}}{100} \times \frac{\text{Contract Sum}}{\text{Contract Period}} \times \text{Period of delay}$$

Hudson's formula found favour with the judge in *Ellis-Don* v. *Parking Authority of Toronto* (1978) 28 BLR 98. In this case, the judge stated that neither counsel before him had been able to think of a better approach.

Emden's formula

This formula can be found in *Emden's Building Contracts and Practice, eighth edition, Volume 2* (page N/46) by Bickford-Smith. The formula is identical to the *Hudson* formula, save that the head office overheads percentage (and profit) used in the formula is the actual percentage based on the contractor's accounts and is arrived at as follows:

$$\text{H.O. (profit) percentage} = \frac{\text{Total Overhead Cost (Profit)}}{\text{Total turnover}} \times 100$$

Emden's formula was approved in the case of *Whittall Builders Company Ltd* v. *Chester-le-Street District Council* (1985) – unreported. The judge clearly stated the principles behind *Emden's* formula as follows:

> 'What has to be calculated here is the contribution to off-site overheads and profit which the contractor might reasonably have expected to earn with these resources if not deprived of them. The percentage to be taken for overheads and profits for this purpose is not therefore the percentage allowed by the contractor in compiling the price for this particular contract, which may have been larger or smaller than his usual percentage, and may not have been realised. It is not that percentage (i.e the tendered percentage) that one has to take for this purpose but the average percentage earned by the contractor on his turnover as shown by the contractor's accounts.'

In *J. F. Finnegan* v. *Sheffield City Council* (1989) 43 BLR 124, the judge endorsed *Emden's* Formula as follows:

'I infinitely prefer the Hudson Formula which in my judgement is the right one to apply in this case, that is to say, overhead and profit percentage based upon fair annual average, multiplied by the contracts sum and the period of delay in weeks, divided by the contract period.'

Note – The judge referred to the *Hudson* formula, when in fact it ought to have been *Emden's* formula.

Eichleay's formula

A similar formula to *Emden's* formula was developed by *Eichleay* in the United States in *The Appeal of Eichleay Corporation*, ASBCA 5183, 60–2 BCA (CCH) 2688 (1960) and this has found approval in the US courts, *Capital Electric Company* v. *United States* (*infra*). This formula uses the actual overheads (and profit) in a similar manner to *Emden*, but the total value of all certificates (the final contract price, including remeasurement and variations) is inserted in lieu of the contract sum.

The logic behind the use of a formula is shown in Figure 5.13.

Line a-a represents the contractor's anticipated or actual head office overheads (depending upon the formula used). Line b-b represents the contractor's anticipated turnover on all projects. Profile c-c represents the contractor's anticipated turnover on the present project. Profile d-d represents the contractor's actual turnover on the present (delayed) project. Profile e-e represents the contractor's actual turnover on all projects.

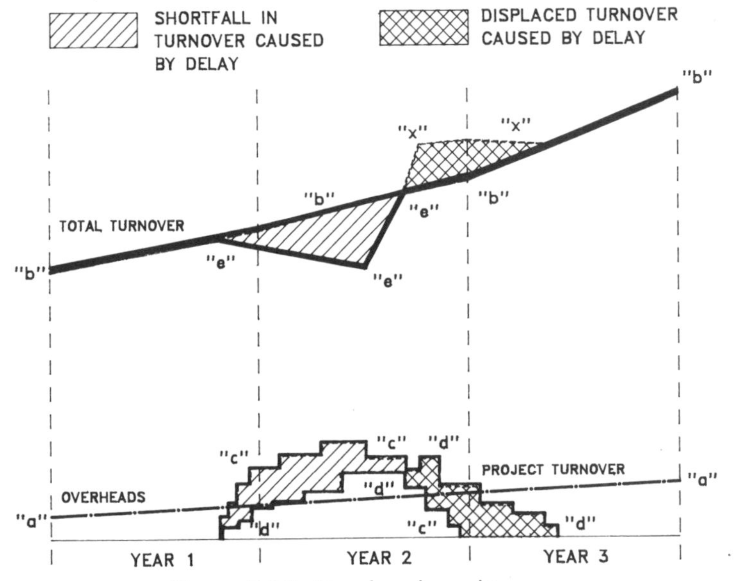

Figure 5.13 Overheads and turnover

It will be seen that the delay has caused the actual turnover on the project (d-d) in the early months of the project to be considerably less than would have been the case if there had been no delay. Accordingly, the total actual turnover (e-e) has fallen below anticipated level (b-b). During the latter months of the project, the actual turnover on the present project (d-d) continues during the period of prolongation (making up for the shortfall in the earlier months). In theory, the actual turnover on all projects during the period of prolongation should increase (see x-x) because the turnover on the delayed project in the latter months was not included in the planned turnover for the same period. However, this increase can only be achieved if the resources on the present delayed project can be released to generate more work on a new project. Unless the contractor can take on more resources, it will have to forego new work which it could otherwise have taken. Therefore, as a result of the shortfall in turnover during the delay, the contractor is unable to recover sufficient overheads from the delayed project to make the requisite contribution to its total overheads.

The various formulae used will enable the contractor to calculate the loss of contribution to its head office overheads as a result of the delay. As the contractor has been unable to release his resources to earn the contribution to overheads on another project, he must earn a similar contribution by making a claim on the delayed project.

It will not normally be necessary for the contractor to submit a graphical representation of its turnover and overheads in the above manner as the use of formulae are well known. Where there is resistance to the use of a formula, illustrations using actual data may be persuasive.

However, when a project goes seriously wrong, the use of a formula may produce a substantial underestimate of the costs of prolongation. A contractor may have to increase the time spent by its managerial and supervisory staff of its head office to cope with the particular problems of the project. Numerous variations and other delaying matters may place greater demands on managerial staff including purchasing, planning, costing, quantity surveying and administration staff. It may be necessary to place a director, in a full time role, to deal with the overall management of the project (where none would have been necessary if the project had gone according to plan).

Before leaving overheads, it is worthwhile considering the different circumstances between the *Tate & Lyle* case and those cases where a formula was accepted as a fair means of calculating overheads to be reimbursed.

In the *Tate & Lyle* case, the court was considering the cost of managerial time spent on work done to remedy an actionable wrong. It had nothing to do with a delayed project. In the cases which approved the use of a formula, the courts were concerned not only with the cost of managing a project which was delayed, but they were also considering the *loss of productivity* (loss of contribution) of the contractor's overhead resources. That is to say, because of the delay, the managerial time could not be used to earn the required contribution to overheads on the delayed project, nor could it be used to earn the required contribution from other existing projects (as this would mean recovering additional expense from other employers who were not in default) or additional projects (which could not be undertaken on account of key resources being retained on the delayed project). With the greatest respect, the circumstances of the *Tate & Lyle* case are sufficiently distinguishable from most cases involving delay and there appears to be strong grounds to resist any suggestion that this case places doubt on the use of an appropriate formula (subject, of course, to reasonable evidence and the circumstances applicable to the delayed project).

Profit

The principles behind a claim for loss of profit arising out of a delayed contract are similar to those applicable to a claim for overheads. It should be noted that some contractual provisions only provide for recovery of additional cost or expense. Where that is the case, a claim for loss of profit is not permissible under the terms of the contract. However, unless there are clear terms to limit the contractor's remedy to those contained in the contract (that is, excluding a common law claim), the contractor may be able to make a claim for loss of profit under the general law. The JCT forms of contract permit reimbursement of loss of profit.

Having established that there is a contractual, or common law, right to recover profit lost as a result of delay, what level of profit is reasonable and what standard of evidence to support a claim for loss of profit is required?

It is an impossible task to show that, save for the delay, the contractor would have been successful when tendering for a particular project (which he declined, or submitted a deliberately high bid) and that, having been awarded the contract for the project, he would have made a profit on it. If that was the appropriate test, no claim for loss of profit would succeed.

However, it may be necessary for the contractor to show some evidence that he was given the opportunity to tender for other projects and that he could not reasonably take advantage of these opportunities because of the fact that his resources were retained on the delayed project. In formulating a claim for loss of profit, the contractor would be advised to keep a record of the following:

- All tenders submitted and awarded (so that a success ratio can be established);
- All projects for which the contractor was invited to tender, but which

were declined or a deliberately high tender submitted (this may cover a period of several months before the present delayed project has overrun, since decisions to decline new work may have to be taken in advance as soon as the overrun is anticipated);

The former is relatively easy to illustrate. The latter may need some analysis to establish that any bids were deliberately high. This should be possible by a bid ratio technique (a system of recording the nett cost included in each tender as a percentage, or factor, of the successful tender).

Example

Nett cost for constructing a project = C, say £100 000

Successful tender sum = T, say £105 000

Bid Ratio = T/C = £105 000/£100 000 = 1.05

Any tenders with a bid ratio above an established competitive bid ratio would qualify for deliberately high pricing. This technique may require statistical analysis and adjustment for 'rogue' bids and errors.

Other evidence, such as proximity of the submitted tender to the competitive range of other tenders, may suffice. Further, a general analysis of construction activity during the period of overrun may be acceptable. Limitations on the contractor's bonding facility may also be a factor.

If the contractor can demonstrate that, on the balance of probability, he would have been able to obtain other contracts during the period of overrun, that alone ought to be sufficient to establish the claim in principle. In a United States case, the employer, the United States Government, contended that the contractor was required to prove that he was capable of taking on the extra work which he alleged was lost as a result of the government's delay and that he could have made a profit on it. It was held that the contractor had produced unrebutted evidence that he could not have taken on any large construction jobs during the various delay periods due to the uncertainty of delays and limitation on his bonding capacity. The mere showing of these facts is sufficient to transfer to the government the burden of proof that the contractor suffered no loss or should have suffered no loss, *Capital Electric Company* v. *U.S.* (Appeal No 88/965, 7.2.84) 729 F 2d 743 (1984).

A very simple approach was adopted in *Whittall Builders* v. *Chester-le-Street (supra)*. The judge was satisfied that there was sufficient activity in the construction industry at the relevant time that it was reasonable to assume that *Whittall* would have been able to obtain other profitable work.

Hudson, Emden or _Eichleay_? Percentage to be used: period for calculating the relevant percentage

A great deal will depend on the nature of the delay. If the sole reason for a particular delay is extra, or additional work, contemplated by the variation clause in the contract, it may be appropriate to use *Hudson's* formula (see Variations – *infra*). If the reason for delay is breach of contract, or if periods of delay caused by variations cannot be disentangled from periods of delay caused by breaches of contract, it is suggested that the remedy is by way of damages, in which case *Emden's* formula is appropriate.

At tender stage, the contractor will be looking at historical data (based on several years expenditure on overheads and the recorded turnover for the same periods). Some adjustment may be made for anticipated changes in turnover in the future overheads. In any event, the percentage for overheads in the contractor's tender should be a realistic estimate of the probable apportionment of overheads in the rates for the work in the contract. The level of profit in the tender may have no relationship whatsoever to historical data, but it will depend on the profit (or loss) which the contractor anticipates should be allowed, having regard to external market factors and operating turnover requirements. Where a positive profit has been allowed in the tender, and where there has been no substantial change in the market, the *Hudson* formula may be fair to both parties where delay is caused by variations.

Where a negative profit has been allowed in the tender, adjustment to the percentage may be considered, particularly if the delay is out of proportion to the value of additional work and/or there had been an improvement in the market (part *Hudson*, part *Emden*). Where the delay was not unreasonable, having regard to the value of variations, adjustment for overheads only (ignoring the negative profit percentage) may be the applicable solution. This would depend on the terms of the contract and the circumstances of the case.

Where a formula is used, there may be some difficulty in deciding upon the appropriate period to be taken for establishing the turnover and overheads and profit in the formula (see Figure 5.14).

Period 'a' (prior to commencement with possible adjustment for anticipated changes) represents the period used for *Hudson's* formula.

Period 'b' (the original contract period) represents the period used for *Eichleay's* formula (see *Construction Contracts: Principles and Policies in Tort and Contract* by I. N. Duncan Wallace at page 128). However, period 'c' (the extended contract period) would appear to be equally appropriate.

Period 'd' (prior to commencement of the qualifying delay) would appear to be the most appropriate for *Emden's* formula, since it is the most contemporary period before the percentage is distorted by the qualifying delay (which would normally reduce turnover and increase the percentage for overheads).

Period 'e' (the period of the qualifying delay) would normally be too short for useful figures to be obtained and it would suffer from greater distortion than period 'd'.

Period 'f' (from commencement of the qualifying delay until completion) may be appropriate in certain circumstances but may be subject to distortion.

Period 'g' (period of overrun) is most suitable for the loss of profit element (since this is the period in which the profit ought to have been earned on a new project). However, it is normally too short. Profit from the nearest year's accounts may be appropriate as a basis of assessment.

Contractors may seek to use the period which gives the most favourable result. In practice, the nearest accounting periods which include period 'd' are likely to be the appropriate periods for calculating the percentage for overheads, whilst the nearest accounting periods which include period 'e' are likely to be the appropriate periods for calculating loss of profit. However, since the use of a formula does not purport to produce an accurate result, it is suggested that period (c) should be appropriate (for overheads and profit) in most cases. If claims are to be settled prior to such information being available, the most recent accounting periods may have to suffice.

The accounting periods will not usually coincide with the actual period, in which case an adjustment may be made. For example, assuming that 'c' has been agreed as the appropriate period, the percentage overheads and profit may be calculated as follows;

	Year 1	Year 2	Year 3	Total
Turnover	£1 800 000	£2 000 000	£2 400 000	
	× 8/12	× 12/12	× 4/12	
	£1 200 000	£2 000 000	£800 000	£4 000 000
Overheads	£240 000	£300 000	£300 000	
and profit	× 8/12	× 12/12	× 4/12	
	£160 000	£300 000	£100 000	£560 000
% on and profit	13.33%	15.00%	12.50%	14.00%

A more accurate assessment may be made by graphical means or by using monthly or quarterly figures.

One pitfall when using actual audited accounts is that they may not include any (or the correct) provision in them for the recovery to be realised by payment of the claim on the delayed contract (and possibly other contracts). Provisions in previous years' accounts may have been under or over-estimated and amounts received in the years used for calculation may

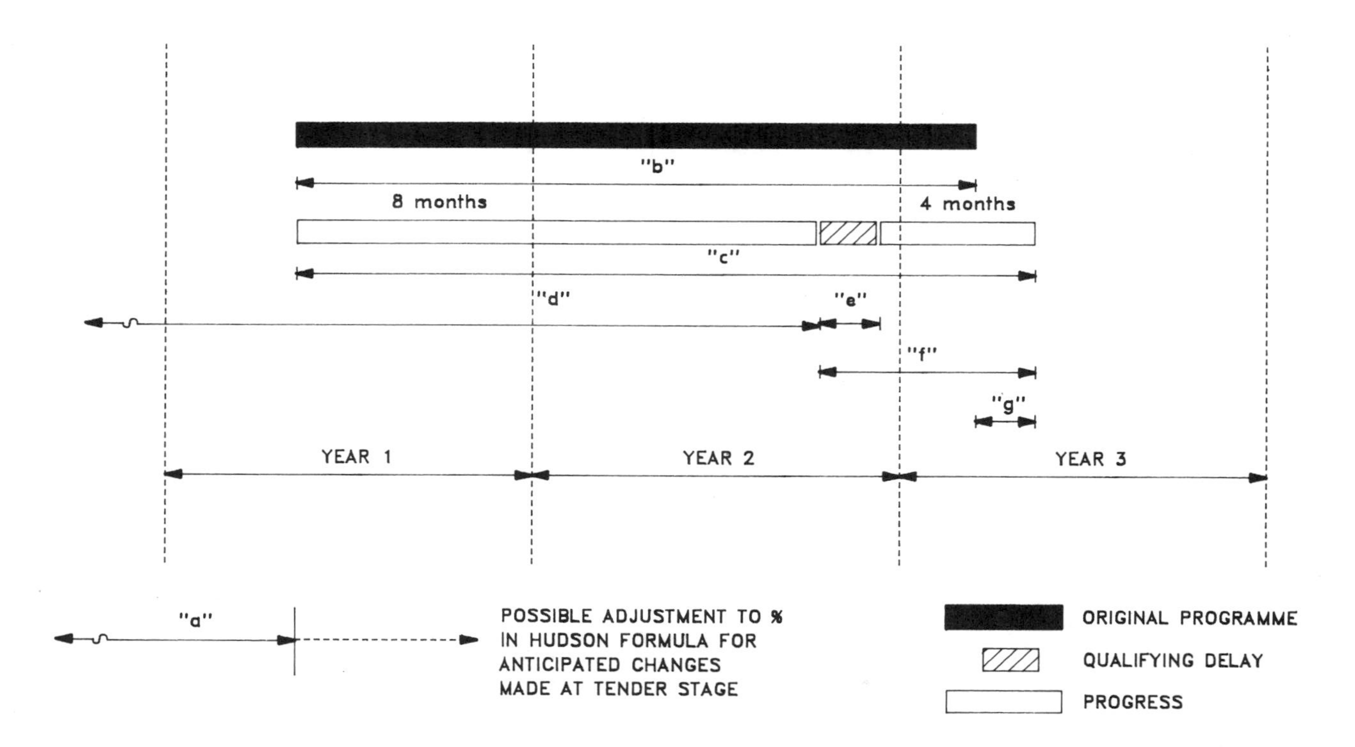

Figure 5.14 Periods for calculating percentage overheads and profit

distort the real figures. Adjustment may be possible if good management accounts are kept. However, unless there are unusual circumstances, it is suggested that these factors will be self-compensating in the long term.

It has been said that a formula produces a result which includes overheads and profit on the overheads and profit included in the contract sum. However, this is not the case if the overheads and profit are expressed as a percentage of the turnover income (and not annual cost), as can be seen from the following example:

Annual cost of all projects	= £60 000
Overheads and profit	= £5 000
Annual turnover	= £65 000
Overheads and profit	= 8.333 % of cost or 7.692 % of turnover
Contract sum of delayed project	= £345 000
Less overheads and profit (7.692 %)	= £26 537
Cost of delayed project	= £318 463
Original contract period	= 300 days
Period of delay	= 70 days

Overheads and profit during period of delay (using contract sum and overheads and profit as percentage of turnover income in the formula)

$$= \frac{7.692}{100} \times \frac{£345\ 000}{300 \text{ days}} \times 70 \text{ days} = £6192$$

Overheads and profit during period of delay (using contract cost and overheads and profit as percentage of annual cost in the formula)

$$= \frac{8.333}{100} \times \frac{£318\ 463}{300 \text{ days}} \times 70 \text{ days} = £6192$$

This example illustrates that there is no mathematical problem when the percentage for overheads and profit included in the tender is the same as the average percentage for overheads and profit on all projects. Adjustment may be necessary if different percentages are evident (as will almost certainly be the case using *Emden's* formula). If this is so, it is a simple matter to convert the percentages so that they are expressed as a percentage of cost, in which case the formula becomes:

$$\frac{\text{Overheads \%}}{100} \times \frac{\text{Contract Cost}}{\text{Contract period}} \times \text{Period of delay}$$

In most cases the traditional use of the formula will be sufficiently accurate. Only where there is a significant difference between average profit and the profit on the delayed project will any adjustment be necessary.

A formula may also produce a suspect result (over-recovery) if the delay being considered is at the end of a project, when most of the work has been done and few key resources are retained on site. The opposite (under-recovery) may occur when the delay takes place during the peak months and the maximum resources are on site. All of the resources should earn a contribution to the overheads and this can be catered for by sensible adjustments to the formula. For example, the following factor may be suitable in some circumstances:

$$F = \frac{\text{Value of work done per day during period of delay on contract}}{\text{Average value of work done per day during total contract period}}$$

Amount of overheads (and profit) = Normal formula result × F

An alternative would be to examine total costs of all projects, the cost of the delayed project and actual overheads during the period of delay (similar to *Eichleay*). This could be ascertained by monthly records. For an example (see also Figure 5.15):

Total cost of all projects, March and April = £160 000
Total head office overheads, March and April = £12 000
Cost of delayed project, March and April = £30 000

$$\text{Overheads percentage} = \frac{£12\ 000}{£160\ 000} \times 100 = 7.50\ \%$$

Overheads allocated to delayed project
during March and April = £30 000 × 7.5 % = £2250

$$\text{Overheads during 45 days delay} = £2250 \times \frac{45}{61} = £1660$$

Problems occur when the cause of delay is a suspension order which applies to the whole, or a substantial part of the works. It is self-evident that the above method would produce a result of zero if all of the works were suspended and no costs were allocated to the project. Nevertheless, *fixed* head office resources would have to be covered by a contribution from the delayed project. It is possible that no management time would in fact be spent on the delayed project. However, this does not mean that more *effective* management time is spent on other projects. Management resources would not be expended on the delayed project (so, in theory, there would be no cost which could be allocated to the delayed projects) thereby making it impossible to justify a claim based on costs as required in *Tate & Lyle* v. *GLC* (*supra*). It must be reasonable to argue that the loss of contribution to overheads should be recovered from the delayed project on the grounds that the contractor's head office resources could not earn the shortfall in contribution on any other project.

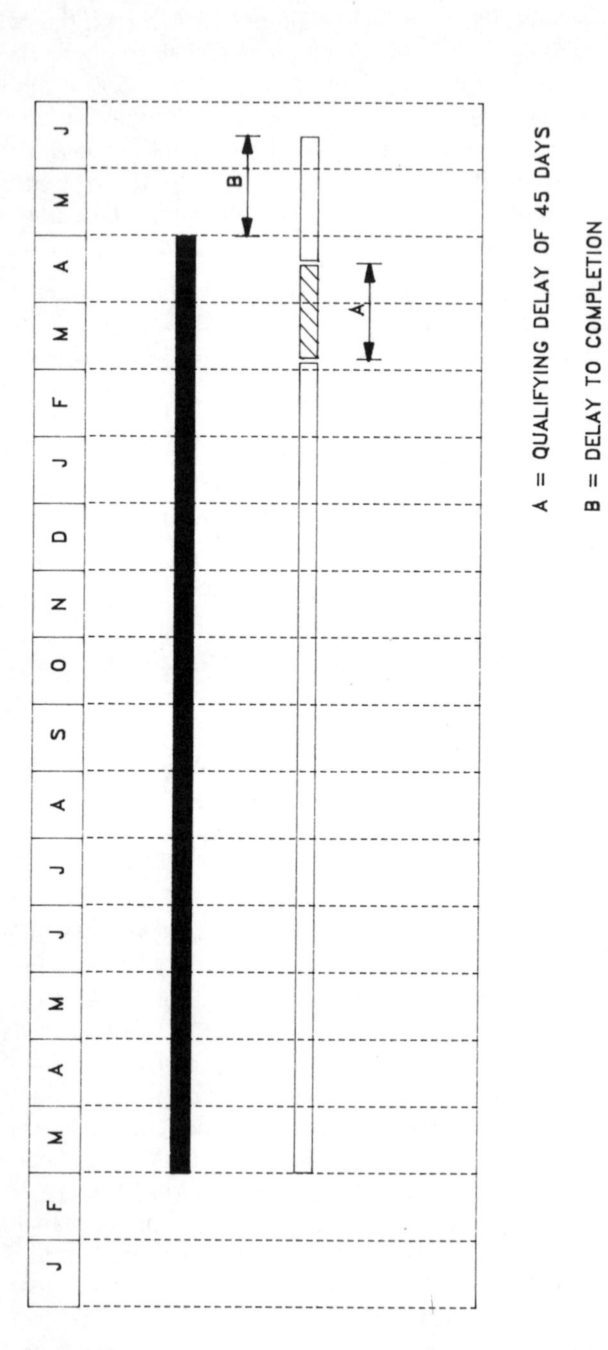

Figure 5.15 Overheads and profit based on monthly accounts during period of delay

Numerous variations to the recognised formulae may be appropriate. In *Finnegan* v. *Sheffield City Council* (*supra*), the contractor argued (unsuccessfully) that the percentage to be used in the formula should be based on a notional contract and the contractor's direct labour cost (excluding subcontractors).

In summary, it is suggested that, unless there are compelling reasons to modify one of the formulae, no adjustment should be necessary when calculating the loss of contribution to overheads (and profit). In most cases, *Emden's* formula, or *Eichleay's* formula are preferable to *Hudson's* formula.

Adjustment for overheads and profit in variations

Many practitioners argue that any recovery of overheads and profit in variations should be deducted from the overheads and profit included in a claim for prolongation. This may be the case in the event of all of the variations being the cause of all of the period of delay. It may not be the case where some (or all) of the variations can be executed within the contract period or they do not cause delay. (See also *The Presentation and Settlement of Contractors' Claims* by Geoffrey Trickey at pp 127, 128).

For example, if variations were executed during a period when there was no delay, the contractor would be paid for them at rates which would include additional overheads and profit. If the contract was to complete on time, no adjustment would be made (but see Variations – *infra*). Therefore, if (after completion of all varied work) there should be delay for another reason (such as suspension), the overheads and profit recovered for this delay (using a formula) would be the appropriate measure of damages for the period of suspension and should stand on its own without adjustment for the overheads and profit recovered in the variations. Similarly, if variations are executed concurrently with other recoverable delays, if it can be shown that they could have been incorporated within the contractor's programme (in the event that the other recoverable delays did not occur) then they may also be discounted and no adjustment made.

In short, any variations which do not cause the delay which is the subject of the prolongation claim may be ignored when making any adjustment for overheads and profit. Conversely, if a variation is the cause of a claim for prolongation, an adjustment should be made.

However, if *Emden's* formula has been used to calculate the overheads and profit during the period of prolongation, the percentage to be used in the adjustment may not be the same as that used in the formula. It should be that percentage which was included in the contractor's tender.

Adjustment for non-recoverable delays

Some delays, such as exceptionally adverse weather conditions, do not qualify or additional payment. Where such delays occur in isolation, it is a simple matter to ignore the period of delay in any calculation of prolongation costs (see Figure 5.16). Where such delays occur in parallel with recoverable delays, reimbursement will depend on the particular circumstances of the case (see concurrent delays – *infra*).

It should be remembered that where a contractor has been forced into a period of adverse weather by a variation, or other qualifying recoverable delay, it may be entitled to reimbursement (*Fairweather* v. *London Borough of Wandsworth – supra*). In these circumstances the adverse weather conditions need not be exceptional in order to qualify for an extension of time and additional payment.

Concurrent delays

A single cause of delay often presents no problem when dealing with prolongation claims. However, in practice, many delays occur at the the same time. Previous examples have illustrated the difficulties which arise when considering extensions of time in such circumstances. The situation is far more complicated when deciding whether, or not, the contractor is entitled to additional payment. There are no easy solutions to the wide variety of practical problems which arise when more than one cause of delay is affecting the progress of the works at the same time. Some delays will qualify for additional payment, whilst others, such as adverse weather conditions (which may qualify for an extension of time) and culpable delay by the contractor will not normally qualify for additional payment.

Contractors are unlikely to offer any concession for concurrent delays when putting forward a claim for prolongation. They cannot be blamed for that (see Negotiations – Chapter 8). The following notes assume that the author of claim is impartial and is attempting to establish what is reasonable reimbursement in the circumstances.

The law applicable to the rights of the parties to damages in the event of concurrent delay is complex. In *Keating on Building Contracts, fifth edition* (pp 193–197), the author discusses the various options which may apply, taking the view that whilst the law appears to be unclear, in the majority of cases, the dominant cause of delay should be the deciding factor. This has been established in cases of exception clauses used in policies of insurance, *Leyland Shipping Company* v. *Norwich Union Fire Insurance Society* [1918] AC 350. It does not appear to be applicable to contracts generally. However, this may sometimes be the case where the facts are clear and the interaction of the various delays are relatively simple to determine.

It is submitted that the 'dominant delay' principle is generally inappropriate for the majority of construction delay claims (with some exceptions). This appears to be supported by the judgement in the *Fairweather* case. If the responsibility for delays can be divided according to the circumstances, apportionment may be appropriate. If it is impossible to disentangle the

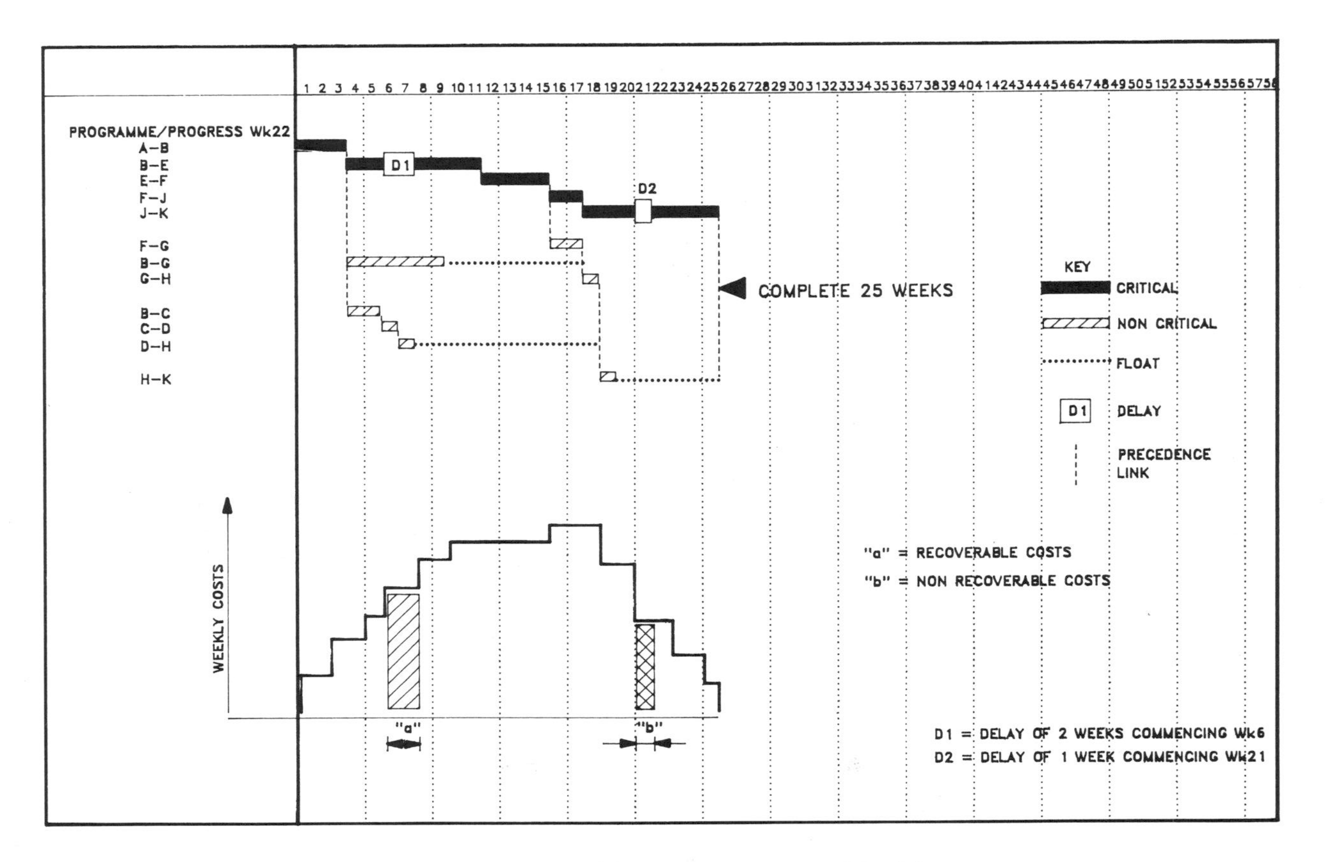

Figure 5.16 Recoverable and non-recoverable delays

causes and effects of the delays, the claim may fail entirely, *Government of Ceylon* v. *Chandris* [1965] 3 All ER 48. If the competing causes of delay are in parallel, only nominal damages may be appropriate, *Carslogie S.S. Co.* v. *Norwegian Government* [1952] AC 292.

The following guidelines may be applicable in circumstances where more than one delay is affecting the progress of the works during the same period of time:

- Where the non-recoverable delay is on the critical path and the qualifying recoverable delay is non-critical, no reimbursement should be permitted;
- Where the non-recoverable delay is non-critical and the qualifying recoverable delay is on the critical path, reimbursement should normally be permitted;
- Where both (qualifying and non-qualifying) delays are critical, then so far as they are of the same duration, no reimbursement should normally be permitted.
- Where a qualifying recoverable delay occurs first, followed by a non-qualifying delay (both delays being on the same or parallel critical paths – see Figure 5.17), there is an argument to support the view that reimbursement should be permitted.
- Where a non-recoverable delay occurs first, followed by a qualifying recoverable delay (both delays being on the same or parallel critical paths), there are grounds to argue that no reimbursement should be permitted.

There may be circumstances which merit a departure from the above guidelines. For example, the greater part of the contractor's management and supervisory staff may have been retained on site to deal with a complex variation which has caused a delay of lesser duration than a concurrent period of exceptionally inclement weather. If it can be shown that the contractor's staff could have been released at an earlier date (had there been no variation), then reimbursement may be permitted notwithstanding the concurrent non-recoverable delay.

The above guidelines should not affect the contractor's rights to recover time-related costs which are exclusively in connection with an activity which has been delayed by the employer (such as the cost of supervisory staff wholly employed on the section of work which which has been delayed by the employer).

Delayed release of retention

When a project is delayed, the certificates which release the retention held by the employer are also delayed. The delay in issuance of the necessary certificates will give rise to a claim for finance charges on the retentions for

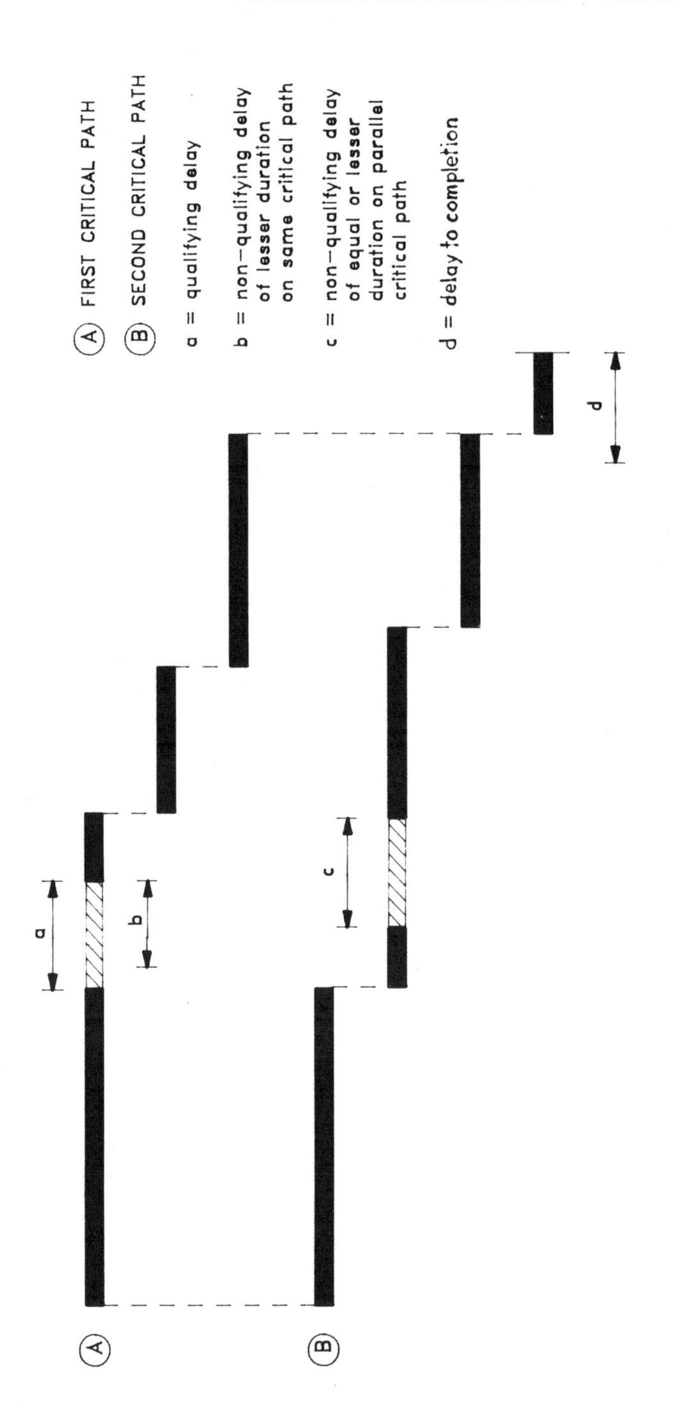

Figure 5.17 Concurrent delay; qualifying delay occurring first

the period of delay. Allowance will have to be made for non-recoverable delays.

5.9 Disruption and loss of productivity

In order to illustrate the effects of disruption and/or loss of productivity it may be necessary to establish that a planned orderly timing and sequence of events was affected by causes within the employer's control to the extent that the contractor was prevented from carrying out the work in the planned orderly timing and sequence. The planned sequence may not be that which was envisaged at tender stage. The project manager may have planned an alternative sequence and this should be the basis of comparison. It may not be necessary to show that there was delay to any activity or that the completion date has been delayed.

Much has been written about the contractor's rights to additional payment in the event of delay when the contractor's programme shows early completion, *Glenlion* v. *Guinness Trust, supra.* Whilst this issue was not decided, the judge referred to two authorities of importance:

'In regard to claims based on delay, litigious contractors frequently supplied to architects or engineers at an early stage in the work highly optimistic programmes showing completion a considerable time ahead of the contract date. These documents are then used (a) to justify allegations that the information or possession has been supplied late and (b) to increase the alleged period of delay, or to make a delay claim possible where the contract completion date has not in the event been extended.' *Hudson's Building and Engineering Contracts, 10th edition* at page 603, and:

'...Sometimes contractors at the commencement of or early in the course of a contract prepare and submit to the architect a programme of works showing completion at a date materially before the contract date. The architect approves the programme. It is then argued that the contractor has a claim for damages for failure by the architects to issue instructions at times necessary to comply with the programme. *Whilst every case must depend upon the particular express terms and circumstances,* it is thought that the contractors' argument is bad;...' (emphasis added), *Keating on Building Contracts, fourth Edition, first Supplement.*

Example

If, for example, the delay of five weeks on bar D (see Figure 5.4) was caused by a suspension order issued immediately upon commencement of the works, the contractor would be entitled to claim the non-productive costs of its site establishment and overheads during the period of delay. These costs would not have been incurred (or they would have been productive costs) if the suspension order had not been issued. Similarly, if the delay of four weeks on bar E (see Figure 5.4) was caused by a variation, the time related costs and any disruptive element of cost would be recoverable as part of the value of the variation. These arguments are valid whether, or not, the delays caused the completion date to be extended. These problems appear to have been contemplated by the judge at page 104 of the report, 'It is unclear how the variation provisions would have applied.'

Whilst the majority of costs claimed are likely to be time-related, they are claimed for disruption rather than prolongation. The *Glenlion* case does not appear to affect the contractor's rights to claim in the appropriate circumstances.

Many disruption claims fail because the contractor is unable to show that the additional costs were caused as a result of matters for which the employer was responsible. In some circumstances, it may be possible to compare actual productivity during a period of disruption with productivity during a period when no disruption was evident. An example of this method was used by the judge in *Whittall Builders Company Ltd* v. *Chester-le-Street District Council (supra)*. In this case, the contractor was able to illustrate that the average productivity was £108 per man-week during the period of disruption and that the average productivity was £162 per man week during the period after the breaches of contract which caused the disruption had ceased (see Figure 5.18). The loss of productivity was therefore one third.

This example was applicable to disruption on the project as a whole and where the nature of the work carried out each month was similar. Where only part of a project is affected, it may be possible to record productivity before, during and after the disruption affecting that part of the works.

Comparison between actual productivity and the allowance in the tender may not be appropriate as a basis of calculation. This method does not take into account errors in the tender. Further, the project team may have changed the method of construction assumed by the estimator. What needs to be considered is the actual productivity with that which ought to have been achieved using the proposed method and sequence that the contractor would have used if there had been no disruption.

In many circumstances, it is difficult or impossible to calculate the cost of disruption of each individual element. A global approach may be the only solution, *Crosby* v. *Portland Urban District Council, (supra* – Chapter 1). This method may be appropriate where the evidence of delay and disruption is overwhelming and there is no significant default on the part of the contractor. If it can be shown that the contractor was partly responsible for the disruption, this type of claim may fail entirely, or the additional costs may have to be borne, in part, by the contractor.

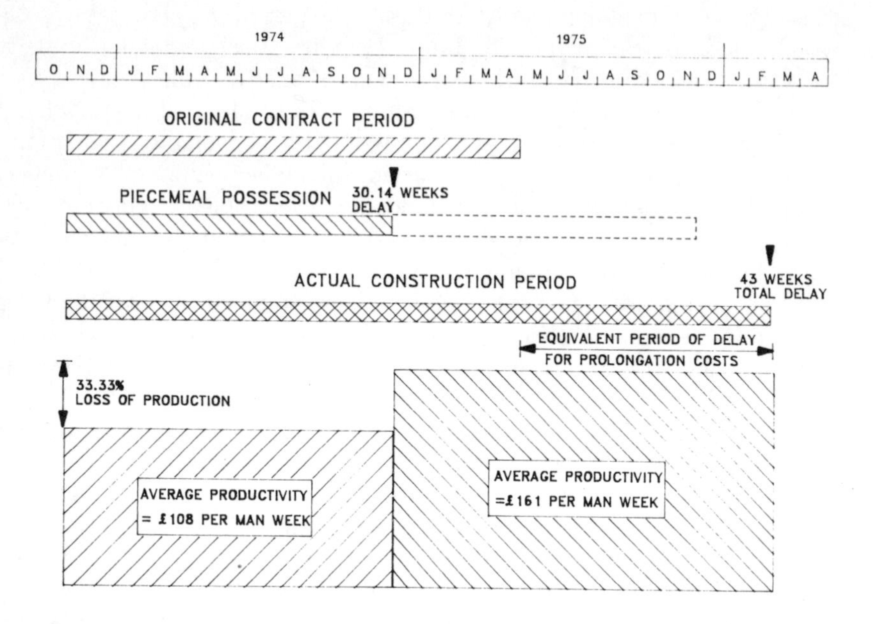

Figure 5.18 Whittall Builders Company Ltd *v.* Chester-le-Street District Council 16.5.1985

5.10 Claims for acceleration

In the event of delay to the progress of the works, the employer, or the contractor, may be faced with deciding whether, or not, there are good grounds to accelerate the progress of the works to bring about earlier completion (to the whole, or part of the works).

From the employer's point of view, acceleration may be advantageous in the following circumstances:

- Where it is essential to achieve completion by an earlier date for commercial reasons;
- Where the delays qualify for additional payment, there is a real probability that the cost of acceleration will be less than the cost of prolongation for the period which can be reduced by acceleration;
- Where there may be substantial savings in escalation costs as a result of earlier completion;
- Where the actual loss to the employer for late completion is greater than the liquidated damages which may be recovered from the contractor.

Some forms of contract (for example GC/Works/1 Edition 3) provide for acceleration. However, the contractor's consent is usually required and the acceleration cost is normally agreed beforehand. Where there are no contractual provisions, a separate agreement will be required. In any event, the terms of an acceleration agreement (including matters required to be dealt with pursuant to clause 38(2)(e) of GC/Works/1) should contain provisions in the event of:

- Subsequent delay by qualifying events which would entitle the contractor to an extension of time for completion (thereby delaying the earlier date for completion);
- Failure to complete by the earlier completion date for reasons which do not qualify for extensions of time (the employer may wish to increase the rate of liquidated damages in the light of his revised anticipated loss).

Whatever the reason for acceleration (even if the contractor is partly responsible for delay and is already liable for liquidated damages), the contractor is likely to be in a strong bargaining position when terms are agreed. The employer should be reasonably confident that the objectives of an acceleration agreement will be met before concluding any deal.

From the contractor's point of view, acceleration may be advantageous if he is in culpable delay and the cost of acceleration is less than the cost of prolongation.

However, when a contract is delayed and no (or insufficient) extensions of time have been made, the contractor may be faced with a dilemma. Should the contractor proceed to complete later than the completion date and run the risk of liquidated damages or should he accelerate the progress of the works to eliminate or reduce that risk?

Very often, pressure is brought to bear on the contractor to improve progress. The language used in these circumstances usually avoids the term 'accelerate', but the contractor is intended to be left in no doubt that he is being pressed to take measures to improve the progress of the works. Veiled, or patently open, threats of deducting liquidated damages may sometimes be used. The contractor's options are:

- To keep his nerve in the belief that the extensions of time will eventually follow (or be awarded in arbitration), or
- To take all of the necessary measures to improve progress and bring about earlier completion, or
- To take some measures to improve progress in the hope that some extension may subsequently be made to the actual completion date.

The decision to accelerate in such circumstances is not easy. If the contractor has a 'cast iron' case for extensions of time, then the first option is probably the best. In these circumstances, the right to recovery of acceleration costs may be in doubt. If the architect, or engineer, has responded to all requests for an extension of time giving reasons for not making an extension, or explaining why an extension was for a lesser period than the contractor's estimate, the contractor is better placed to judge whether, or not, the extension is reasonable or capable of being reviewed. However, if there is no response, or if the response is an unreasoned rejection of the contractor's application for an extension of time, the contractor has no means by which to judge the eventual outcome which may result from further representations. All of these circumstances, including the pressure which may be brought to bear to improve progress, will influence the contractor's decision to accelerate.

Where it can be shown that the contractor was entitled to an extension of time when he took the decision to accelerate, and that the architect, or engineer, ought reasonably to have made the extension of time promptly, there are grounds to argue that the contractor is entitled to reimbursement of reasonable acceleration costs. The claim will be based on the premise that there was a breach of contract (that is, failure to operate the extension of time provisions). The success of such an argument will depend on:

- Whether the contractor had complied with the contractual provisions to give notice and particulars of the delay in accordance with the contract;
- Whether the architect, or engineer, had properly considered all of the circumstances and events for each delay before making, or rejecting, an application for an extension of time (there may be a considerable difference between a genuine attempt to make an extension where the conclusion was merely wrong, and a rejection out of hand without proper, or any, consideration being given to the matter);
- To what extent the contractor had communicated his intention to accelerate and the circumstances at the time of making the decision;
- Whether, or not, the contractor's decision was a sensible commercial decision in the circumstances;
- Whether, or not, the contractor's claim for the costs of acceleration were less than the probable cost of prolongation (it may be equitable to reimburse the contractor for the costs of acceleration if the employer was ultimately going to benefit by a saving in the amount of the contractor's probable claim for prolongation – that is to say that the employer should not benefit from his own default – *Alghussein* v. *Eton College* – Chapter 1, *supra*).

Invariably, it can be shown that the reason for failing to make extensions of time was a result of pressure from the employer on the architect, or engineer. Sometimes this is evident from the conduct of the employer's representatives and the professional team at meetings (or even in correspondence). Where this is not evident, it may come to light during discovery of documents or upon cross-examination in arbitration or litigation. Unfortunately, it is becoming increasingly common for some powerful employers to use the threat of termination of services (or the promise of future work) as a lever to put pressure on, or influence the architect or engineer.

If such pressure or influence was present, the contractor would have a prima facie claim for reimbursement (see *Morrison-Knudsen* v. *B.C.Hydro & Power* and *Nash Dredging* v. *Kestrell Marine Ltd* – Chapter 1, *supra*).

If it should be established that there is a case for reimbursement of acceleration costs, there is the difficult task of proving the actual amount of the claim. Costs which need to be considered are:

- Non-productive overtime – That is, the premium rates paid to operatives for working outside of normal hours. Not all of the overtime hours are recoverable. Only those hours in addition to the allowance in the contractor's tender should be claimed (if the contractor had always planned to work nine hours per day and Saturday mornings in order to complete within the original contract period, he could only claim the additional hours in a claim for acceleration);
- Additional cost of employing extra staff and operatives – Higher rates of pay, incentives, travelling time, subsistence and transportation costs of importing labour;
- Loss of productivity – An increase in the number of staff and operatives does not necessarily bring with it a proportional increase in production. On a congested site, labour cannot be utilised as efficiently. The co-ordination of various activities and trades becomes more demanding and there is likely to be a greater incidence of waiting time between activities;
- Increase in the use of lighting and power – Inevitable in winter and in large buildings and basements;
- Increase in the hire of equipment and plant (sometimes fuel only).

Whatever the reasons for acceleration, the contractor ought to be aware, before incurring the additional costs, that care should be taken to keep good records to enable the above costs to be substantiated. It should also be borne in mind that, whatever the moral grounds justifying acceleration, in practice, this head of claim is one of the most difficult to justify on legal grounds.

5.11 Variations

Variations to the works are almost inevitable. Therefore, all standard forms of contract contain provisions to deal with them. Some variations can be made without affecting the progress of the work and with no change in the method, sequence and cost of the work to be done in the variation. In such circumstances, the rates applicable to the contract can be applied to the measured quantity of work in order to arrive at the value of the variation. However, even when these simple rules are applied, there may be some indirect costs which need to be addressed.

For example, if the cost of insurance premiums have been included in the 'Preliminaries' sections of the bills of quantities, there may have to be an adjustment made to the 'value related' element of the insurance premiums in the bills to reflect any change caused by variations. Where there is a decrease in the contract price as a result of variations, there may be no adjustment to the cost of insuring the works (depending upon the insurers' practice in this regard). However, a decrease in the contract price may justify a reduction in the allowance for employers' liability insurance. Likewise, if small tools and equipment are priced in the preliminaries section of the bills, an increase may be justified if the contract price is increased by variations. Where there is a decrease in the contract price, the likelihood of the contractor being able to save on the amount of tools and equipment is remote (unless the reduction in work was known well in advance of the need for the necessary tools and equipment).

In practice, most variations have some effect on the progress of the works and the method of executing the work. Where it is possible, each variation should be valued taking into account all of the delaying and disruptive elements which are directly related to the variation. Common factors which affect the valuation of variations are:

- Changed conditions or circumstances – The varied work may be carried out in different circumstances than those contemplated at tender stage for reasons which are entirely related to the nature of the variation itself. For example, the contractor may have allowed for excavation to reduced levels using scrapers to deposit spoil in a temporary spoil heap for future disposal. Due to a variation to add a length of surface water drain across the site in the location of the spoil heap, the contractor is forced to excavate and load into lorries and cart away most of the spoil in one operation. The revised method takes longer so that more work is done in wet weather and the operation is more costly. There is no delay or disruption to the works as a whole. This change could, and should, be dealt with by valuation under the variation provisions in the contract. There is express provision for such an eventuality in clause 13.5.5 of JCT80.
- Changed quantities – Some changes in quantities have a significant effect on cost, even when the nature of the work and the method of executing the work is unchanged. For example, an increase in the volume of concrete may require working overtime in order to complete a floor slab which may be critical to the activity planned to commence the following day.

Another example is where an increase in quantities causes some of the work to be carried out later. If the quantity of brickwork increased by twenty per cent, and using the same resources, the time to execute the work (but not any other activities or the contract as a whole) was extended into another pay increase, then the extra costs resulting from the pay increase should be reflected in the value of the variation (assuming a fixed price contract).

- Changed timing – Work of a similar nature to that contained in the contract may be ordered at different times so that material and labour costs are not the same as those for the original work.
- Small quantities – Variations requiring ordering and execution of similar work in small quantities may involve loss of purchasing discounts and increased prices payable to subcontractors who may have to return to site after completion of the original subcontract work.
- Time-related costs – Where it is possible to isolate a period of delay to part, or the whole, of the works to a single variation (or group of variations), the time-related costs may be reflected in the value of the variation. For example, a major variation to the ground floor structure may cause the time taken to reach completion of the first floor slab to be delayed by one week. It may be appropriate to include the costs of the entire concrete, steelwork and carpenter resources, including concrete mixers, pumps, dumpers, tower-crane, supervision and other preliminary items in the value of the variation. Additional time may be required as a result of actual remeasured quantities exceeding the quantities in the contract bills.

Time related costs were the subject of a dispute under conditions of contract which were similar to those contained in clause 52 of the FIDIC and ICE conditions of contract. In *Mitsui Construction Co Ltd* v. *The Attorney General of Hong Kong* (1986) 33 BLR 1, the executed work in a tunnelling contract was significantly different to that measured in the bills of quantities. The changes in quantity were not a result of a variation order given by the engineer. The contract period was twenty-four months. The result was that the contractor had taken much longer to complete the works and the engineer had granted an extension of time of 784 days. The contractor argued that he was entitled to compensation for the costs of the extra time taken to complete the works. The employer argued that the contract did not empower the engineer to agree or fix any adjusted rates. The Privy Council ruled that the engineer was empowered to vary the rates, thereby opening the way to take account of the time-related costs in the valuation of the variation. It should be noted that clause 2.2.2.2 of JCT80 contains provisions which would enable time-related costs to be taken into account in the event of a variation arising out of errors in the quantities in the contract bills.

In some circumstances, there may be arguments as to whether the contractual provisions permit the valuation of disruptive, or time-related, elements as part of the variation. The proviso to clause 13.5 of JCT80 is unclear and unhelpful in this regard. It would appear that the rules governing the valuation of variations are sufficiently flexible to permit a very wide interpretation of them so as to enable the quantity surveyor to adopt a sensible approach according to the circumstances. Contractors should bear in mind that it is in their interests to include as much as possible in the valuation of variations so that an element of profit can be recovered on the extra costs. This is particularly important where the provisions of the contract limit reimbursement to cost, or expense, if the additional payment is claimed under any other provisions.

5.12 Dayworks

Payment for work on daywork is usually reserved for circumstances where there is no other reasonable means of valuing the work to be done. Some contracts provide for the contractor to give advanced notice of any work to be done on daywork. There are usually strict time limits for submission of daywork vouchers. It is important to follow the contractual provisions so that the time and materials can be properly recorded and agreed. Contemporary notes setting out the reasons for recording the work on daywork may be helpful. It is important to include all incidentals, such as small tools and transport. Signatures verifying the times and materials used may not signify that payment will be made in the daywork account. However, proper records of such work can be of assistance as supporting documents for other methods of payment.

5.13 Fluctuations

Most fluctuating price contracts use a recognised formula which is applied to the value of work done each month. The base date is predetermined at tender stage and fluctuations are calculated by reference to the published indices each month and the base index. Some contracts contain a 'cut-off date' in the event of delayed completion. However, not all of the effects of price increases may be recovered under the fluctuations clause. If there is a qualifying recoverable delay, any shortfall in recovery which can be substantiated may be included in the contractor's claim for additional payment under the appropriate contract provisions.

In the event of delay during a fixed price contract, work is progressively carried out at later times than allowed for in the tender. The estimator ought to have allowed for the anticipated increases in cost during the contract period in accordance with the tender programme. By comparing actual progress and the value (or cost) of work done each month with anticipated progress and value (or cost) of work in accordance with the programme, it is possible to determine the probable effects of inflation as a result of the delay. The actual monthly value and relevant monthly index can be used to compare the planned monthly value and index as shown in Figure 5.19.

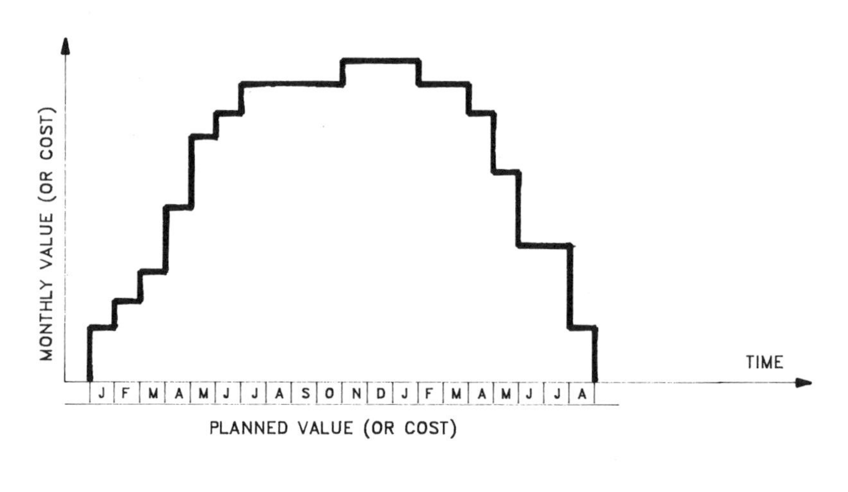

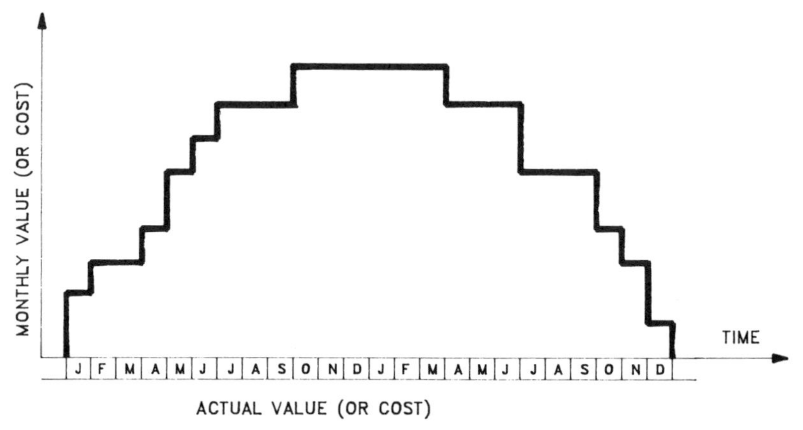

$$\text{INFLATION} = \sum\left(\frac{\text{AMV} \times \text{MI}}{\text{BI}}\right) - \sum\left(\frac{\text{PMV} \times \text{MI}}{\text{BI}}\right)$$

AMV = ACTUAL MONTHLY VALUE (OR COST)
MI = INDEX FOR RELEVANT MONTH
PMV = PLANNED MONTHLY VALUE (OR COST)
BI = BASE INDEX (AT TENDER)

NOTE: IF MONTHLY VALUE IS USED, RESULT MAY HAVE TO BE ADJUSTED FOR PROFIT ELEMENT

Figure 5.19 Calculation of fluctuations using published indices

It should be borne in mind that this method may not be accepted as a means of measuring the additional cost due to the delay. However, providing that suitable adjustments can be made for materials and sub-contracts let at fixed prices (which are not changed during the contract), materials on site and other factors which may be applicable, this method is generally recognised as a reasonable means of calculating reimbursement. Other evidence, such as comparison of actual invoices and wage rates paid at different times may be required.

5.14 *Quantum meruit*

A well drafted variation clause will enable the employer to make substantial changes to the works without invalidating the original contract. Nevertheless, variation clauses do not enable the employer to vary the works without limit. In *Wegan Construction Pty. Ltd.* v. *Wodonga Sewerage Authority* (see Chapter 1, *supra*), substantial changes were made and the contractor claimed payment on a *quantum meruit* basis. The variation clause applicable to this case, in part, is almost identical to the FIDIC conditions of contract, and is sufficiently similar to many other forms of contract to justify a detailed analysis of the case.

Clause 40.1 of the contract contained the following terms:

'Variations Permitted. At any time prior to practical completion the engineer may order the contractor to:

(a) increase, decrease or omit any portion of the work under the contract;
(b) change the character or quality of any material, equipment or work;
(c) change the levels, lines, positions or dimensions of any part of the work under contract;
(d) execute additional work;
(e) vary the programme or the order of the work under the contract;
(f) execute any part of work under the contract outside normal or agreed upon working hours;

and the contractor shall carry out such variation, and be bound by the same conditions, so far as applicable, as if the variation was part of the work under the contract originally included therein.

The extent of all such variations shall not, without the consent of the contractor, be such as to increase the moneys otherwise payable under the contract to the contractor by more than a sum which is the percentage stated in the annexure A of the contract sum, or if not stated, by a reasonable amount.

No variation shall vitiate or invalidate the contract, but the value of all variations shall be taken into account and the moneys otherwise payable under the contract shall be adjusted as provided under cl. 40.4.'

It appears, from the judgement, that no percentage had been inserted in annexure A, and the contract was therefore construed on the basis of the term 'by a reasonable amount'.

In the new plans, excavation was increased by twenty per cent; sewer length was increased from 840 metres to 1 181 metres; manholes from nineteen to twenty-seven, requiring a ninety per cent increase in concrete; house connections had increased from forty-seven to ninety-one and the new design included one hundred and sixty metres of excavation below four metres deep which was not shown on the original plans. The contract price was $30 867.40 and the revised contract price was $43 200.

The contractor argued that the change in design was not a variation permitted by the contract and sought to be released from the contract rates and for payment to be on a *quantum meruit* basis.

Held: In the circumstances the amended plans did not constitute a variation permitted by the original contract.

In practice, where there are very wide variation provisions, and the rules for valuing variations allow for departure from the contract rates, it may be difficult to argue successfully that the works should be valued on a *quantum meruit* basis. There would have to be some compelling reasons which would have made it impossible for the contractor to continue on the basis of the original contract. A substantial increase in the value of work may not, on its own, be sufficient reason to escape from the contract rates.

5.15 Finance charges

In nearly all cases, contractors will allow something in their tender for finance charges on the working capital required to carry out the works. There may not be a positive cash flow until final retention is released. Whatever the contractor's anticipated cash flow, as a general rule, if the value of work increases, the additional financing ought to be recovered in the rates for variations (assuming that the finance costs are allocated throughout the rates for measured work).

However, it is often the case that interim certificates do not reflect the true value of the original contract work including variations. In such circumstances the contractor will be incurring additional finance charges on the under-certified sums. Whilst significant changes have taken place in recent years to compensate contractors for the loss incurred as a result of increased finance charges in cases of default by employers, the commercial reality of the high cost, and potential loss, has not been recognised fully in many modern contracts or in the general law. A claim for finance charges on late, or under-certification, will have to be founded on a contractual provision, or for breach of contract.

In the case of *Morgan Grenfell Ltd* v. *Sunderland Borough Council and*

Seven Seas Dredging (1991) 51 BLR 85, it was held that clause 60(6) of the ICE fifth edition enabled the contractor to claim compound interest on amounts which were included in a statement under clause 60(1) if the engineer failed to certify and it was subsequently found that the amounts ought to have been certified.

Most contracts do not have a provision for interest to be paid in the case of failure to certify (or under-certifying). However, if the facts are clear, and there was sufficient information before the certifier to enable a proper valuation and certificate to be issued for the amount claimed to be due, there may be grounds to argue that interest is payable as a result of a breach of contract.

Where delay and disruption occur, the interest on the cost, or on the loss and/or expense, may be claimed as part of the cost or expense. This was held to be the case in *Rees and Kirby Ltd* v. *Swansea City Council* (1985) 30 BLR 1.

Whilst it it not usually essential to include a statement showing the amount of interest on delay and disruption claims, it is a practice which should be encouraged, if only to prompt the architect or engineer to deal with the matters in the earliest possible interim certificate.

5.16 Cost of preparing the claim

In the vast majority of cases, the cost of preparing the claim is not a recoverable cost. However, there are circumstances in which the cost of preparing claims may be recovered:

- If each claim is prepared by the contractor's staff, as and when they arise during the contract, the salaries and other costs of the staff will usually be included in the site or head office overheads and may therefore be included in the general claim for prolongation;
- If, in spite of all requests for an assessment of the amount of the claim (and provided that the contractor has provided all particulars in accordance with the contract) no assessment is made within a reasonable time (and particularly if it has not been made within the period of final measurement or other specified contractual time frame), the contractor would be justified in preparing his own claim and may be entitled to reimbursement – see *James Longley & Co Ltd* v. *South West Regional Health Authority* (1983) 25 BLR 56 at page 57, 'The costs of preparing a final account may be recovered as damages in a suitable case, eg for breach of an obligation on the part of an employer to provide a final account...'. This may include the contractor's own managerial time (provided that it is not included in overheads), *Tate & Lyle Food Distribution Ltd* v. *GLC* – (*supra*);

- Where certain work is done in connection with preparing a case for arbitration, *James Longley* v. *South West Regional Health Authority, supra*. The cost of preparing unnecessary evidence may not be allowed.

5.17 Assessment and evaluation

Assessment and evaluation of delay and disruption claims will depend on the pricing and accounting policy of the contractor. The following should be established:

The tender

How are the overheads and profit distributed in the tender? Loading rates or preliminaries may merit adjustments to any sums calculated using a formula.

Are all of the site overheads (preliminaries) priced in the preliminaries sections of the bills of quantities? If part, or all, of the preliminaries are included in the rates for measured work, some analysis may have to be done to ascertain the sums to be used as a basis of calculating time-related elements (if it is appropriate to use the contract rates for variation delays). An adjustment may have to be made to account for additional preliminaries recovered in the rates for variations (whilst there are circumstances where no adjustment should be made for overheads and profit recovered in variations, an adjustment will usually be justified for any preliminaries recovered in variations).

Accounting practice

Are head office overheads charged to the project? If so, on what basis? Time records? Percentage allocation? *Ad hoc*? Unusually high allocation of costs may have to be justified.

Are finance charges included in general overheads? If so there may be duplication with separate claims for finance charges. This may be overcome by deducting interest and finance charges from the general overheads and making a separate assessment of the finance costs on the average working capital required for the delayed project (excluding claims).

Having established the above, the assessment and evaluation of the claim can proceed without fear of unnecessary duplication or omission.

It is important that all facts, evidence and data upon which any calculations are based are collected and bound in an annotated appendix to the claim. In the narrative of the claim, the author should have set out the basis of the claim giving reasons for any particular method which has been adopted (such as an explanation as to why a particular formula has been used to calculate overheads and profit and any adjustments which have been made).

It is sometimes helpful, and persuasive, to give financial information in tabular and graphical form. This will facilitate a better understanding of the nature of the contractor's claim and may assist in obtaining an early settlement.

Each head of claim should state the source documents used (referring to the appropriate appendix) and any assumptions made for the purposes of calculation or assessment.

5.18 Summary on presentation of claims for additional payment

Similar guidelines to those given for extensions of time are applicable to claims for additional payment. In spite of the fact that contractors may not be reimbursed for preparing a claim, it is usually in the contractor's interest to do so at the earliest opportunity. The temptation to wait until extensions of time are made before submitting a claim should be resisted unless there is real possibility that this will sour relationships beyond repair. In any event a claim should be prepared (even if not submitted) so that the magnitude of the loss or additional cost can be made available to management. The sooner the opposition are made aware of the amounts which are likely to be claimed, the better the chances that funds will be put aside to meet it.

In addition to the details and particulars mentioned with regard to extensions of time (*supra*), the following may be necessary:

- Details of the effects of any delay or disruption on all activities in parallel and subsequent to the circumstances giving rise to the claim;
- An introduction to the claim giving the contractual provisions under which the claim is being made;
- A summary of notices and particulars given during the contract:
- Diagrammatic illustrations where appropriate;
- References to recognised authorities and case law relied upon;
- Additional, or alternative claims under the general law (if applicable).
- A statement setting out the amount of the claim.

Presentation will depend on the type of claim. If several individual claims are made during the course of the project, these need not necessarily be couched in legal language which is sometimes seen in formal submissions.

5.19 Formal claim submission

If individual claims are dealt with and settled promptly during the contract, a formal submission setting out the contractual basis and detailed analysis of the contractor's rights and entitlements will not be necessary. However, if settlement is not reached on these claims, the contractor is faced with preparing a document which, it is hoped, will lead to an amicable settlement at the earliest possible time. This type of claim submission may take a form almost approaching pleadings for arbitration. Some contractors spend considerable time and effort in negotiations which fail because of the lack of a sound, comprehensive and persuasive submission which sets out the contractor's claim and the basis upon which the claim is made. The sooner a formal submission is made, the earlier a settlement can be reached or proceedings can commence. A formal claim submission will include:

Introduction: contract particulars

Names of the parties; description of the works; details of tender and acceptance; the form of contract and any amendments thereto; the contract sum; dates for commencement and completion; phased completion (if applicable); liquidated damages for delay; the programme.

Summary of facts

Date of commencement and practical completion; dates of sectional or partial completion (if applicable); summary of applications for extensions of time; extensions of time awarded; summary of claims submitted; final account and claims assessed (if any); amount of latest certificate and retention; payments received; liquidated damages deducted (if applicable).

Basis of claim

Contract provisions relied upon; common law provisions; contractual analysis and explanation of the basis of the claim.

Details of claim

Full details of every matter which is the subject of the claim. Each separate issue should be carefully set out in a logical format. Key dates, events, causes and effects, references to relevant documents and the like should form the basis of a narrative which fully describes the history of the project and the effects on progress, cost and completion. It is important to distinguish between the causes and effects of delay (and/or disruption), extensions of time and the financial effects of delay and/or disruption. Wherever possible, diagrams, programmes, tables and the like should be included in the narrative (or in an appendix). The extensive use of schedules can be invaluable.

Evaluation of claim

Each head of claim should be calculated, step by step, with explanations and reasons for the methods adopted. Supporting source documents (from which financial data has been used in the evaluation of the claim) should be given in an appendix, or listed, so that the recipient may examine such documents at the contractor's office when considering the claim.

Statement of claim

A brief statement setting out the claimant's alleged entitlements and relief sought, such as extensions of time; sums claimed; repayment of liquidated and ascertained damages (if applicable).

Appendices

Copies of all documents referred to in the claim; programmes; diagrams: schedules; financial data.

6 Subcontractors

6.1 Subcontracting generally

An increasing number of contractors do less work by direct labour and they rely to a great extent on subcontractors for the execution of the work. It is perhaps for this reason (at least in part) that contractors are sometimes unable to provide adequate particulars and substantiation in support of their claims.

At tender stage, contractors may rely on subcontractors' quotations for large sections of the works. The tender may be based on the lowest of all the subcontractors' quotations. Once the contract has been awarded, the contractor will then seek to get better quotations (by negotiation with the original tendering subcontractors or by looking for alternative quotations).

In many cases, the contractor will not award the various subcontracts until it is necessary to do so. For example, the subcontract for painting may not be awarded until a few weeks before the painting is due to commence. The contractor runs the risk of price increases in these circumstances. If there has been delay to the project, prior to placing the order for painting, it will be difficult for the contractor to establish a claim for an increase in the cost of the work. Is the increase in the subcontract price due to the delay to the project, or is the market for painting buoyant at the time of subcontracting (whereas it may have been depressed at the time of tender)? If the painting had been ordered at tender stage, the subcontractor may well have had a claim for increased costs due to executing the work at a later date, but this would have been determined by contractual provisions based on conditions at tender stage.

This practice makes it difficult for the contractor to justify a claim for additional payment. The subcontractor will have no interest in providing particulars (because the extra cost is in his price). The employer will not expect to reimburse the contractor for the extra cost caused by a buoyant market. Nevertheless, the contractor may have grounds for a claim.

If all subcontracts were placed at tender stage, based on the same programme and other contractual provisions, the contractor ought to be able to deal with subcontractors' claims as if they were his own (subject to the practical difficulty of getting subcontractors to give the same notices and particulars to the contractor as the contractor is required to give under the principal contract). In practice, subcontracts are placed progressively during the course of the project. If delays occur throughout the project, as the magnitude of the cumulative delay increases, various subcontracts will be placed on different programmes and base costs. Very often subcontracts will be placed when the contractor's current programme is out of date (sometimes the programme may be obsolete to the extent that the programme shows completion of the subcontract works before the date of placing the order for the subcontract). These problems are not imaginary. They occur regularly in real life and are a constant source of contractual disputes.

It is often a problem to establish the subcontractor's obligations regarding progress and completion of the subcontract works when the order, or subcontract, states that the subcontract works shall be carried out 'in accordance with the contractor's programme'. Which programme? Was it the programme which was in existence at the time of making the subcontract (even if the programme shows the subcontract works to be complete before the time of the subcontract)? Is it to be the next revision of the programme? Is it to be any future revision of the programme? What is the situation if the contractor never produces a revised programme?

The dangers which may arise from the above practices are:

- The period for completion of the subcontract works may be impossible to determine from the subcontract documents, in which case the subcontractor may have an obligation to complete within a reasonable time. A reasonable time for the subcontractor may not be within the time allowed for the principal contract;
- The subcontractor may take on board the obligation to execute the works in accordance with any programme of the contractor.

Even more uncertain and onerous provisions (from the subcontractor's point of view) arise when the terms of the subcontract require the subcontractor to proceed with the subcontract works in accordance with the contractor's reasonable requirements. In the case of *Martin Grant & Co Ltd* v. *Sir Lindsay Parkinson & Co Ltd* (1984) 29 BLR 31, the subcontract contained the following terms;

'2. The Sub-Contractor will provide all materials labour plant scaffolding in addition to that provided by the Contractor for his own requirements haulage and temporary works and *do and perform all the obligations and agreements imposed upon or undertaken by the Contractor under the Principal Contract* in connection with the said works to the satisfaction of Contractor and of the Architect or Engineer under the Principal Contract (hereinafter called "the Architect") *at such time or times and in such manner as the Contractor shall direct or require* and observe and perform the terms and conditions of the Principal Contract so far as the same are applicable to the subject matter of this contract as fully as if the same had been herein set forth at length and as if he were the Contractor under the Principal Contract.

3. The Sub-Contractor *shall proceed with the said works expeditiously and punctually to the requirements of the Contractor and so as not to hinder hamper or delay the work or the portions of the work at such times as the Contractor shall require having reference to the progress or conditions of the Main Works and shall complete the whole of the said works to the satisfaction of the Contractor* and of the Architect and in accordance with the requirements of the local and other authorities.' (emphasis added).'

The works under the principal contract were delayed and the subcontractor was retained on site for a considerably longer period dictated by the progress of the principal contract. The subcontractor contended that there was an implied term that the contractor would make sufficient work available to enable the subcontractor to maintain reasonable and economic progress and that the contractor would not hinder or prevent the subcontractor in the execution of the subcontract works. The subcontractor's claim failed and he was unable to recover the extra costs arising as a result of working on site for a much longer period.

Some of these problems can be avoided by using one of the standard forms of contract which are tailor-made for use with the appropriate principal contract. Some contractors have their own 'look-alike' forms of contract which resemble the standard forms of subcontract but which contain onerous provisions. Subcontractors should not assume that onerous provisions can be defeated by implied terms.

6.2 Nominated subcontractors

Nominated subcontractors have been used in building contracts for over one hundred years. They appeared in the RIBA Model Form of Contract at the beginning of the century. They have a useful and important function where the employer has a genuine requirement to select a subcontractor to execute specialist work. However, the provisions and procedures surrounding their selection and use have become unnecessarily complicated. PC Sums (Prime Cost Sums) in contracts are intended for work to be done by nominated subcontractors or for materials or goods to be supplied by nominated suppliers.

In general, it is better to limit nominated subcontractors to a minimum, and then only for work which cannot reasonably be included in the contractor's own scope of work. Some of the reasons which may justify the use of nominated subcontractors are:

- Where the subcontractor is to undertake design responsibility and the features of the subcontractor's design must be co-ordinated with the principal design of the works;
- Where it is essential to appoint a nominated subcontractor before appointment of the contractor for the principal contract (for example, there may be long delivery periods for plant and equipment to be provided by the subcontractor);
- Where the subcontract works is an extension of work done previously by a particular subcontractor and the same equipment and standards are required to be used in the new works;
- Where the subcontract works are the main requirements of the employer and the building, or civil works, are secondary (for example, in process plants);
- Where the employer, or its designers, have a particular preference for a subcontractor based on previous performance and standard of work.

Having regard to the increasing amount of sophisticated mechanical and electrical installations, including lifts, escalators, heating and ventilating and air conditioning (HVAC), building automation systems (BAS), security systems (such as closed circuit television – CCTV) and a host of new additions to the field of building services, it is not surprising to find these in the form of PC sums which, in total, may make up more than fifty per cent of the total building cost. In these circumstances, if PC sums are used properly, it may be appropriate to nominate subcontractors to do this type of work.

In this context, 'used properly' means that, for a lump sum contract (such as JCT80), the scope of the works to be done by nominated subcontractors should be fully defined at tender stage (of the principal contract). That is to say, the design of the subcontract works should be complete in all of the essential details so that the tendering contractors can appreciate the magnitude, complexity, sequence of other work and any other limitations on their own methods and sequence of working to ensure completion of the principal works by the contract completion date. It is wholly insufficient to describe the works intended to be covered by a PC sum in one or two lines in the bills of quantities, or specification, giving an approximate sum as a guide to the contractor for pricing his attendance and profit.

Quite apart from being contemplated on contractual grounds, it is sound commonsense to completely develop the design of all of the specialist subcontract work alongside the design of the building structure and building envelope. If this is not done, how can the design be co-ordinated to ensure that all of the service pipes, ducts, cable trays and equipment can be built into the spaces allocated for them? It is this lack of co-ordination which leads to conflicts in the services during construction on site and in some cases renders it impossible to incorporate them in the space allowed. This may require late variations to re-route some of the services in unsightly bulkheads and lowered ceilings. In extreme cases, valuable floor space may have to be sacrificed or, if it is not too late, storey heights may have to be increased. The 'knock-on effect' may include redesign of curtain walls and substantial changes to lift cables, controls and machinery. The cost of all vertical components and finishes will increase.

These direct costs may be a small proportion of the costs of delay and disruption and may cause substantial loss of revenue for the employer. Consultants who embark upon a design up to tender stage without taking account of these potential problems may find themselves being sued by the employer who has not had his building on time and has paid considerable additional sums of money to the contractor for the privilege.

These problems arise when the contract contains PC sums which are no better than provisional sums in disguise. If, for example, the design of the kitchen equipment is not complete, or not capable of being adequately defined, at tender stage, a provisional sum should be used in preference to a PC sum. If PC sums are used for work which is really provisional, the design team may be misleading the contractor and the problems which arise may be costly to resolve. The work which is eventually ordered under a PC sum may be considerably more complex than could reasonably be contemplated at tender stage. Is the subcontract works (as ordered) the same as the original intention, or is it a variation? A variation to the principal works may not be a variation to the subcontract works (because the 'baseline' for design may not be the same for the principal contract and the subcontract). If a detail is issued during the progress of the subcontract works, the contractor may be justified in claiming an extension of time and additional payment (on the grounds that it is a variation to the original design), whereas the subcontractor was aware of the new detail and had allowed for it in its price and programme.

Many of these problems can be avoided by careful planning and co-ordination of design by the employer's professional advisers, so that the contractor is left in no doubt, at tender stage, what is contemplated in the work which will be done by nominated subcontractors.

6.3 Contractors' rights to object to nominees

Most forms of contract contain provisions for the contractor to object to any nominee on limited grounds (clause 35.4.1 of JCT80 and clauses 59A.(1) and 59(1) of the ICE fifth and sixth editions respectively). JCT80 contains detailed provisions and alternative procedures which may apply. However, in general, the contractor will have a right to object to a nominated subcontractor for the following reasons:

- If the subcontractor will not enter into a subcontract on terms containing provisions which indemnify the contractor against the same liabilities as those for which the contractor is liable to indemnify the employer and which indemnify the contractor against any claims arising out of default or negligence of the subcontractor;
- If the subcontractor shall not agree to complete the subcontract works in accordance with the reasonable directions of the contractor and to enable the contractor to discharge its obligations under the principal contract;
- If the subcontractor will not agree to complete the subcontract works within the period specified in the proposed subcontract;
- If there are reasonable grounds for the contractor to believe that the subcontractor is unsuitable or is financially unsound.

The first three reasons are usually catered for in standard forms of subcontract designed to operate alongside the appropriate standard form of principal contract. Any attempt by the contractor to impose more onerous provisions will usually be thwarted by predetermined tender procedures which are known by the contractor (such as those contained in JCT80 and the standard form of tender – NSC/1). However, if the principal contract contains amendments and more onerous provisions than the standard form of contract, the contractor would be within his rights to insist on similar provisions in the subcontract, so far as they were applicable to the subcontract works.

The third reason may arise if nomination procedures are not followed, or if the nomination is made during a delayed project. If there has been no delay and the period for completion contemplated by the subcontractor is inconsistent with the contractor's original programme, the contractor will have a prima facie case to object unless the nominee agrees to comply with the programme. If delay has occurred, various problems may arise:

If the contractor is in delay, but no extension is justified, the contractor may reprogramme the remaining work to allow a shorter period for work to be done by a subcontractor to be nominated at a future date. For example the contractor may cause delay of two weeks to activity B-E (see Figure 6.1). The contractor's revised programme may show a reduction in the period allowed for activity J-K which is for work to be done by a nominated subcontractor (see Figure 6.2) so that the completion date is preserved. Activities B-E and J-K are on the critical path but none of the other activities are critical.

Is it reasonable for the contractor to object if the nominee can complete within the original period allowed, but refuses to agree to a shorter period? Can this be overcome by making an extension of time so that the subcontractor can be accommodated, thereby enabling the contractor to escape liability for liquidated damages for his own delay? On the strict wording of clause 25.3.1 of JCT80, completion of the works must be likely to be delayed by a cause which is a relevant event and as the real cause of delay was the contractor's own default, it may not be possible to make an extension. Is time at large? Is the contractor liable for unliquidated damages? Do the contractual provisions need revision to deal with this situation?

Delays may occur for which extensions of time may be due, but for which no extension has been made. There may be a dispute as to the contractor's entitlement to an extension. If a subsequent nominated subcontractor cannot complete its work by the current completion date, is the contractor justified in objecting to the nominee (even if some of the previous delay was caused by the contractor's own default)? Should an extension be made to accommodate the nominated subcontractor? What is the situation if it should subsequently be found that no extensions of time were justified for delays prior to the date of the nomination? Is the nomination made late (even if the nominee was able and willing to commence work on the day that the contractor would be ready for him to commence work)?

The problems which arise when realistic dates for work to be done by nominated subcontractors are out of synchronisation with the contract completion dates and/or the contractor's programme are common. A commonsense solution may be the only way ahead. Some of these problems have been considered in the courts. The House of Lords heard an appeal in the case of *Percy Bilton Ltd* v. *The Greater London Council* (1982) 20 BLR 1 (HL). A nominated subcontractor withdrew his labour from site on 28 July 1978 and went into liquidation. The subcontractor was behind programme at the time of his withdrawal with some forty weeks of the subcontract period remaining. On 31 July 1978, *Bilton* (the contractor) terminated the subcontractor's employment. The (extended) contract completion date at this time was 9 March 1979. Some of the defaulting subcontractor's work was done by a temporary subcontractor (Home Counties Heating & Plumbing Limited) under architect's instructions and on 14 September, *Bilton* was instructed to enter into a nominated subcontract with a new subcontractor (Crown House Engineering Limited). The new subcontractor withdrew his tender on 16 October and on 31 October *Bilton* was instructed to enter into a nominated subcontract with Home Counties. Negotiations between *Bilton* and Home Counties were concluded on 22 December 1978 on the basis that Home Counties would commence work on 22 January 1979 and that the period for completion of the subcontract works would be approximately fifty-three weeks (complete about 23 January 1980). Various extensions of time were granted, but the architect only granted an extension of fourteen weeks (to 14 June 1979) under clause 23(f) of JCT63 for the delay caused by renomination (see Figure 6.3). Further delays occurred; the contractor completed late and the *GLC* deducted liquidated damages. The contractor contended that time was at large and that liquidated damages could not be deducted. It was held that the delay arising out of the

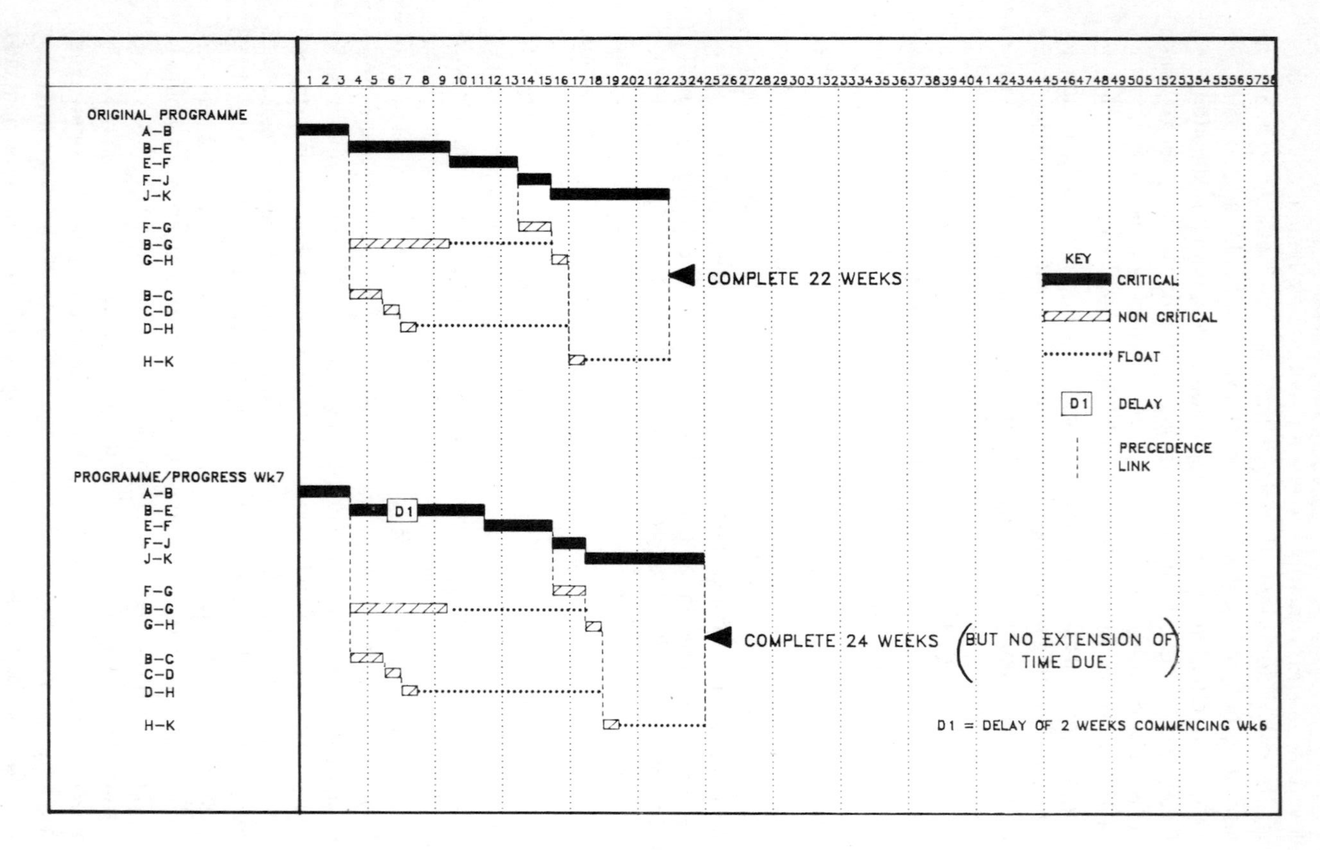

Figure 6.1 Critical delay due to contractor's default

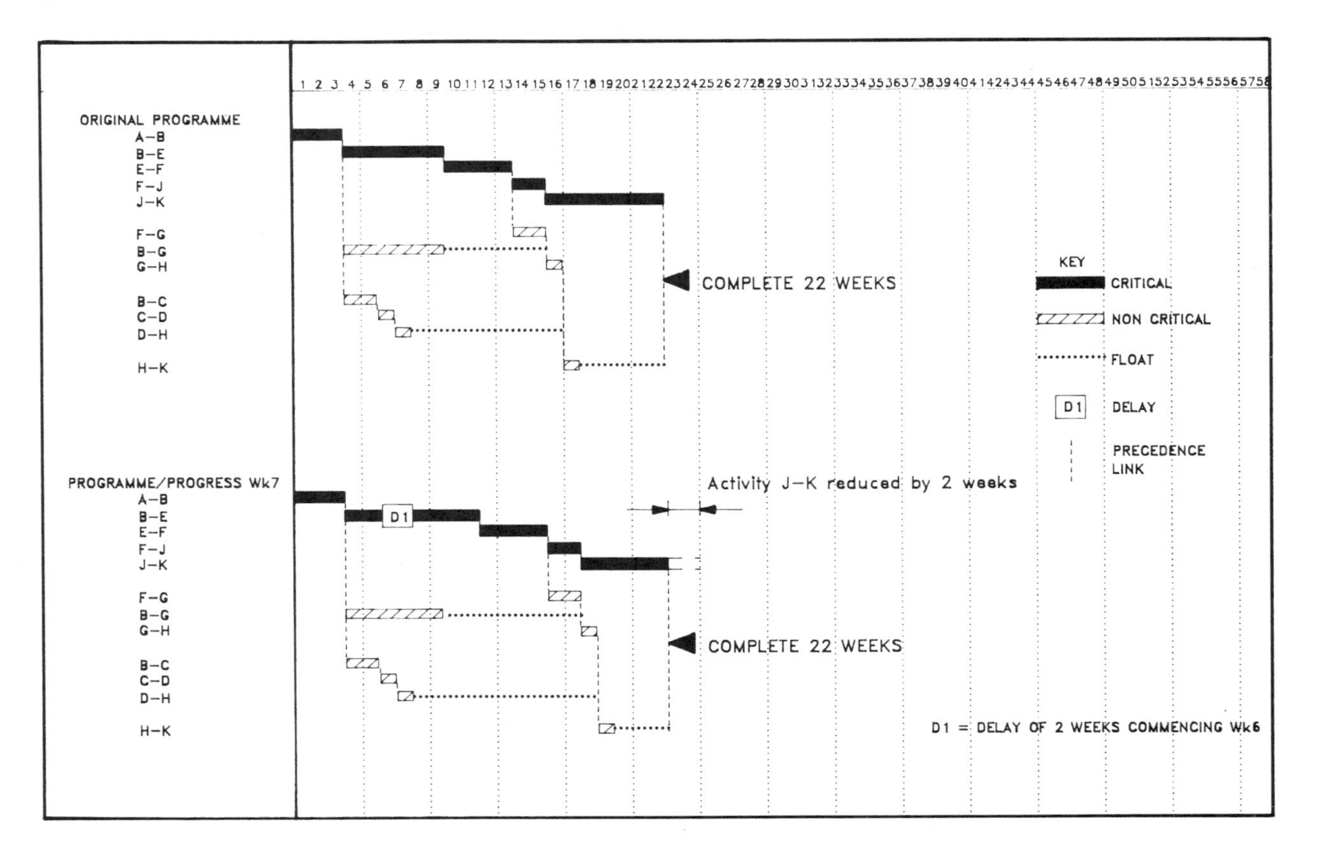

Figure 6.2 Critical delay due to contractor's default – Reduced period for subcontractor to preserve completion date

renomination fell into two parts. The first part was due to the original subcontractor's default and the second part was due to the unreasonable time taken to engage Home Counties to complete the work. No extension of time was justified for the first part of the delay (however,it appears that the extension of time granted by the architect included the first part of the delay), but the architect was empowered to grant an extension of time for the second part of the delay. As the first part of the delay was not due to the employer's default time was not at large and liquidated damages could be deducted.

An important aspect of this case was reported in the Court of Appeal *17 BLR 1* (at page 18):

> 'A quite separate argument by Mr Garland is what is described as his "overshoot" submission; that is to say that, at the time of the application for the re-nomination, the new subcontractor's date for completion was later than the plaintiff's date for completion and that, since this would make it impossible for the plaintiffs both to accept the new subcontractor and to comply with the provision in their own contract as to time for completion, therefore the time provision must go completely, time will be at large and the right to liquidated damages will disappear.
>
> I do not accept this argument. The contractor, faced with a subcontract with such a provision as to completion, would be entitled to refuse to accept the subcontractor under clause 27 [of JCT63]; or what the subcontractor could do would be to say that he would not agree to accept the subcontract unless at the same time the employer would agree to an extension of time for the completion of the main contract.'

The above argument found support in the House of Lords, *20 BLR 1* (at page 15).

It should be noted that this case dealt with *renomination* which was not due to the employer's default. If these circumstances arose with respect to the original nomination of a subcontractor to execute the work covered by a PC sum, the result would probably be very different. The contractor may have a claim for breach of contract and/or a claim arising out of a late instruction pursuant to provisions in the contract.

In a similar case of *Fairclough Building Ltd* v. *Rhuddlan Borough Council* (1985) 30 BLR 26, a nominated subcontractor ceased work in September 1977 and the subcontractor's employment was terminated. The subcontractor was eight weeks late at the time of termination. The standard conditions of JCT63 had been amended to exclude delay by a nominated subcontractor (unless such delay was due to a reason for which the contractor could obtain an extension). The original date for completion of the principal contract was 2 May 1977 and an extension of time for strikes occurring prior to the subcontractor's withdrawal from site was granted to 10 May 1978. The architect did not issue an instruction to renominate a new

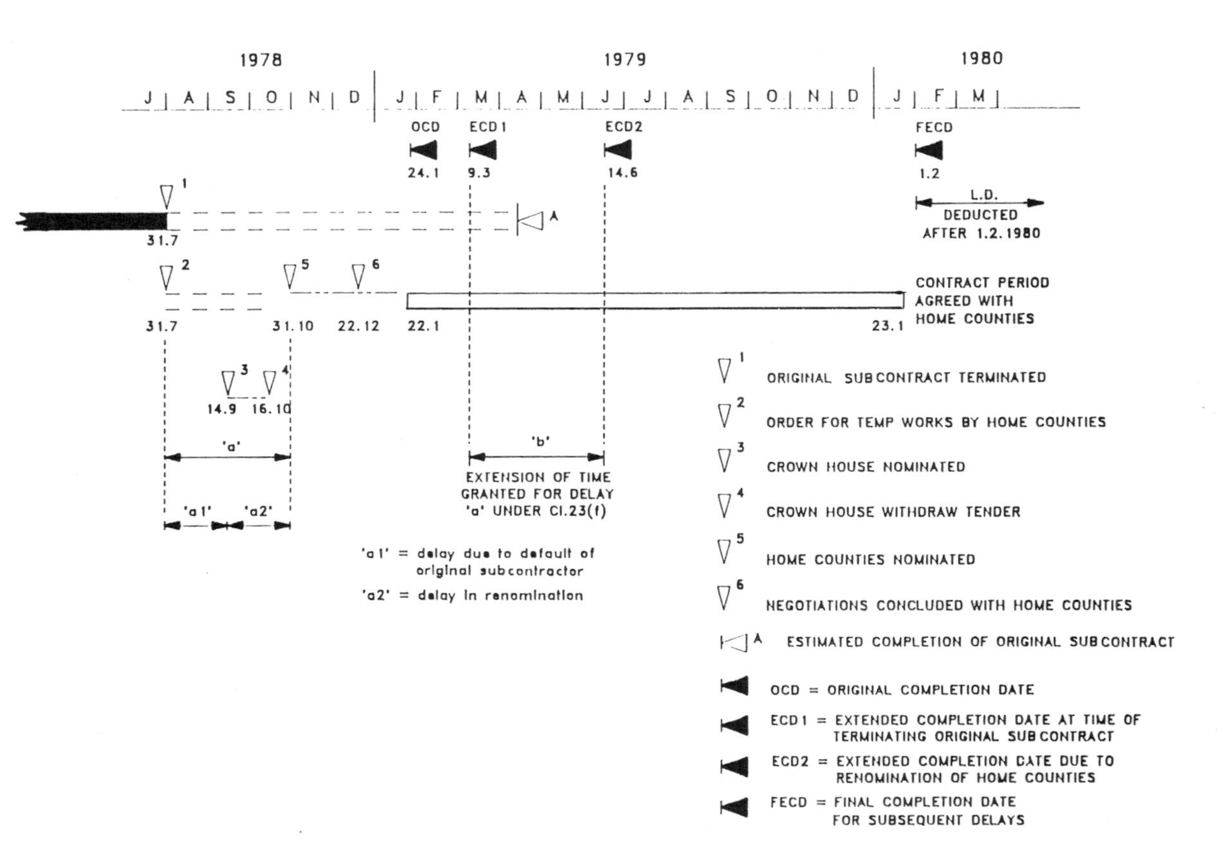

Figure 6.3 Percy Bilton *v.* Greater London Council

subcontractor until 24 February 1978. The contractor objected to the renomination on the grounds that it did not include making good defects in the original subcontract work and that an extension of time would be required to cover the time required by the new subcontractor (twenty-seven weeks from acceptance of tender) which would overrun the date for completion of the main contract (see Figure 6.4). The architect replied (on the latter issue) stating 'I would confirm our intention to grant an extension of time in connection with the re-nominated Sub-contractor's programme time at such time as the effect on your overall programme can be ascertained.'

It was held that the contractor was entitled to refuse the nomination. With respect to extensions of time, the following is of practical importance, *30 BLR 26* (at page 41):

> 'In the present instance delay until 24 February therefore falls on the contractor [following *Bilton* v. *GLC,* but on the grounds that the period taken to renominate by 24 February 1978 was not an unreasonable time]. If, when his contractual completion date is some two and a half months off he is asked to do work which will take six months to complete we see no reason for saying that the contract must be so construed that he cannot insist on an extension of time under the main contract to bring it in line with the proposed subcontract,...'

and at page 42:

> 'It may well be that the doing of such work would not delay *actual* completion of all outstanding work but if the contractor is required on 24 February to do work which cannot be done until September it appears to us at least arguable that he could not be in breach of contract by reason of failure to do that part of the work until September and thus that he is entitled, **if he does not exercise his right to prevent nomination**, to an extension to that date.' (bold emphasis added).

The main difference between the *Bilton* case and the *Fairclough* case was that *Fairclough* had asked for an extension of time to cover the period to complete the work required by the new nominated subcontractor, and the architect had intimated that he would grant an extension of time, whereas no extension had been requested in the *Bilton* case.

Similar problems arise where the contract contemplates the use of named subcontractors to execute work. However, if the contractor is unable to enter into a nominated or named subcontract for reasons which are justified, there may be machinery to overcome some of the difficulties by way of a variation or by omitting the work or by substituting a provisional sum (clauses 3.3.1 of IFC84 and 35.2.3 of JCT80).

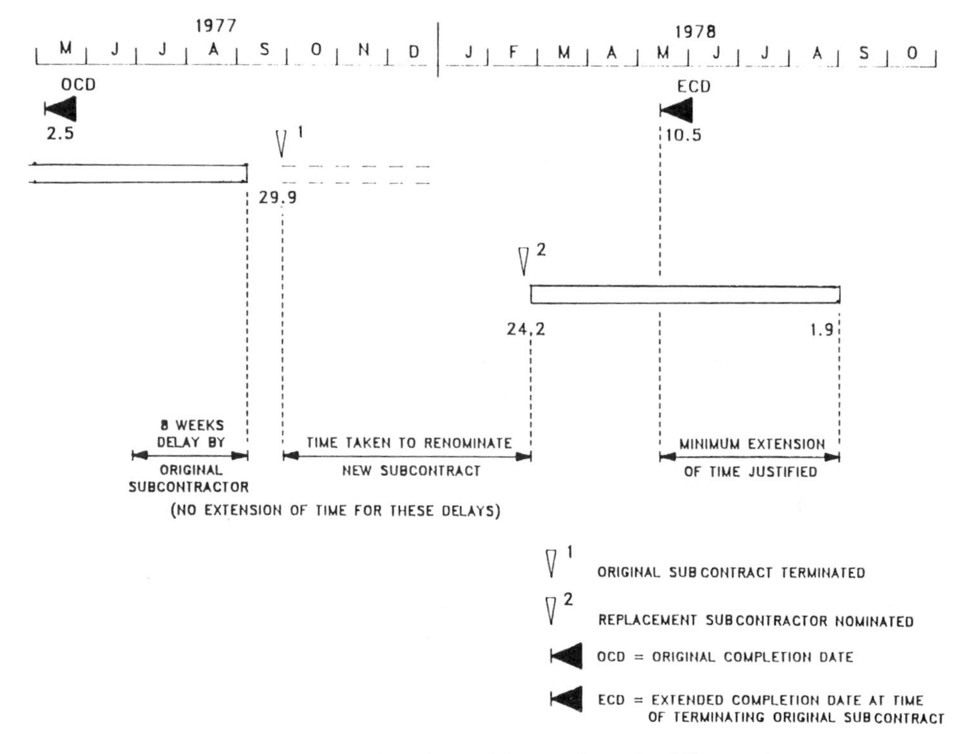

Figure 6.4 Fairclough Building Ltd *v.* Rhuddlan B.C.

6.4 Subcontractors' programmes

In most cases, the contractor's programme will indicate overall periods for work to be done by each subcontractor. The programme may show separately, first, second and final fixing and various sections of the subcontract work. Whatever the level of detail shown on the contractor's programme, many subcontractors will need to subdivide their work into several activities when preparing their own programmes. If the contractor has been given sufficient design information when tendering for the work, he will have been able to prepare his programme taking into account many of the factors which govern the sequence of the subcontractor's work. Assuming that the contractor's programme is still valid (based on progress and the current contractual completion date), the contractor and the subcontractor ought to be able to agree a realistic programme which is consistent with the overall programme. It would be unusual if some minor reprogramming of the principal works and/or the subcontract works was not necessary at the time of subcontracting. A competent contractor, given sufficient information at tender stage, ought to be able to accommodate such reprogramming without raising an objection or subsequent claim.

In some cases, the subcontract works may be on the critical path, in which case the subcontractor's programme and the overall programme need to be given careful attention, preferably before the subcontractors submit their tenders for the subcontract works. This can be facilitated by ensuring that the contractor and all tendering subcontractors have detailed discussions at pre-tender stage. Where the subcontract works are not critical, the subcontract period may be open to negotiation. For example, if the activity B-G in Figure 5.2 (*supra* – Chapter 5) represents work to be done by a subcontractor, the options for the subcontract period may be:

- Commence at the beginning of the fourth week and complete in six weeks (earliest start);
- Commence at the beginning of tenth week and complete by the end of the fifteenth week (latest start);
- Commence at the beginning of the fourth week and complete by the end of the fifteenth week (earliest start and latest finish);
- Any period between the beginning of the fourth week and the end of the fifteenth week (which may be more or less than six weeks duration).

These options may have a bearing on the subcontractor's price for executing the subcontract works and should therefore be discussed before submission of the subcontractor's tender (whether the subcontractor is domestic or nominated). They may also have a bearing on the contractor's attendance (for example, the period required for scaffolding). In the case of a domestic subcontractor, the contractor can use the optimum solution to arrive at the best tender for the main works or (if arising after award of the principal contract) to obtain a saving on its original estimate for the works. In the case of a nominated subcontractor, the employer may enjoy the benefit of the optimum solution.

Another difficulty arises where the subcontract is executed on or about the date of commencement of the main works, but the subcontract works are due to commence several months later. Delays to the main works which occur prior to the date of commencement of the subcontract works may qualify for an extension of time (for completion of the main works). However, the progress of the subcontract works has not been delayed (since the subcontractor has not yet commenced work) and there may be no provision to adjust the completion date of the subcontract works. It is therefore important to make provision in the subcontract for the commencement and completion dates of the subcontract works to be adjusted in such circumstances. This may be overcome by stating a period for completion of the subcontract works and providing for the subcontractor to commence work within a specified period of the contractor's written notice. This may be ideal for contractors, but subcontractors may require provisions to enable them to recover any additional costs which may arise from delayed commencement.

6.5 Extensions of time for completion of subcontract works

Most forms of subcontract contain provisions for extensions of time to be made for the following reasons:

- Delay for which the contractor is entitled to an extension of time for completion of the works pursuant to the principal contract;
- Delay or default on the part of the contractor, or persons for whom the contractor is responsible (such as other subcontractors).

If the subcontract works is on the critical path, a qualifying delay which affects the subcontract works will have equal effect to the completion periods for the subcontract works and the main works. If the subcontract works is not on the critical path, delays which occur may have different effects on the relevant completion dates. For example, delay on the critical path may give rise to an extension of time for completion of the main works, but no extension of time may be necessary for completion of the subcontract works. Alternatively, a qualifying delay to the progress of subcontract works may justify an extension of time for completion of the subcontract works, but no extension may be necessary for completion of the main works (subject to the contractor subsequently needing an extension – see Chapter 5 *supra*).

With the exception of delay on the part of nominated subcontractors under some JCT forms of contract (*infra*) delays by other subcontractors (or by the contractor) may entitle the subcontractor to an extension of time, but the contractor may not be able to obtain an extension of time for completion of the main works. In such circumstances, various claims and counterclaims may arise (see Chapter 7 – *infra*).

6.6 Delay by nominated subcontractors

The JCT forms of contract (JCT63 and JCT80) contain certain provisions which can only be regarded as being against the interests of the employer. JCT80 (clause 25.4.7) provides for extensions of time in the event of delay on the part of nominated subcontractors or nominated suppliers which the contractor has taken all practical steps to avoid or reduce. No doubt contractors have insisted upon this provision in the light of experience and on the grounds that they have not freely had control over the selection of the nominee. However, if the contractor is to be given the opportunity to discuss all essential details with the nominee, prior to nomination, and having regard to the contractor's right to object to any nominee, these provisions should be removed. Before removing these provisions, the employer and its professional team should be prepared to make all nominations in plenty of time for the contractor and the subcontractors to agree to the programme and for orders to be placed so as to prevent delay. If these requirements cannot be met, and the extension of time provision for delay on the part of nominated subcontractors is deleted, contractors will be more likely to exercise their rights to object, thereby causing delay to the progress of the main works. The contractor may also be entitled to an extension of time for completion of the main works pursuant to clause 25.4.6.

Notwithstanding the provisions of clause 25.4.7 of JCT80, the subcontract provisions (clause 11.2.2.1 of NSC/4a) preclude an extension of time for completion of the subcontract works in the event of delay by the subcontractor. The contractor may therefore avoid liability for liquidated damages under the principal contract and the subcontractor may become liable directly to the employer.

6.7 Architect's consent to grant an extension of time to a nominated subcontractor

JCT80 requires the architect's consent to grant an extension of time to nominated subcontractors (clause 35.14). Some architects are reluctant to exercise their powers promptly on the grounds that the contractor may use it to justify an extension of time for completion of the main works. This is not necessarily the case, and these powers should be exercised as soon as possible having regard to the completion periods of the respective subcontract (which may, or may not, be critical to the completion period for the main works – *infra*).

In the case of qualifying delays, an extension of time may, or may not, be necessary for completion of the main works. In the case of delay by the contractor (or other subcontractors), the architect may have an obligation to give his consent to grant an extension of time to a delayed subcontractor. Failure to do so at the appropriate time may provide the delayed subcontractor with grounds to argue that time for completion of the subcontract works became at large.

6.8 Design and drawings provided by the subcontractor

In contracts where the responsibility for design rests with the employer, any design of the subcontract works by the subcontractor is deemed to be the employer's design. Therefore, any delay in design by the subcontractor will be considered to be delay by the employer. However, where the subcontractor is required to provide installation drawings, these may not be considered to be design drawings and the subcontractor will be liable to the contractor for any delay caused by late issuance of installation drawings, *H.Fairweather & Co Ltd* v. *London Borough of Wandsworth* (*supra*).

Difficulties often arise where design and installation drawings are to be provided by the subcontractor. What constitutes a design drawing and what constitutes an installation drawing? There are no reasons why these should not be defined in the principal contract (definitions in the subcontract may be of no consequence since the contractor may argue that such definitions were not part of the principal contract). In the absence of such definitions, it is suggested that the following principles may be applied:

- Design drawings include drawings which require calculation and/or co-ordination with other parts of the works (such as works being designed by other subcontractors);
- Installation drawings include drawings which merely represent the subcontractor's interpretation of the design having regard to all design information provided by the employer's design team.

In the former case, the design of the subcontract works may depend on design development of other parts of the works, for which the employer assumes responsibility for design. The design team will have to ensure that the design of all installations, and the building, fit together. In the latter case, the subcontractor must be given sufficient information on all other installations to enable him to complete his installation drawings.

Some contracts attempt to place responsibility for co-ordination of design by subcontractors (in addition to co-ordination of the installation) upon the contractor, or on the various subcontractors. This is a recipe for disaster and employers should be advised to avoid this practice. It is likely to cause considerable delay and extra cost which, in spite of careful drafting of the contractual provisions, will almost certainly end up being the responsibility of the employer.

6.9 Variations to the subcontract works

Variations to the subcontract works are usually subject to the same treatment as variations to the main works. However, the design of the subcontract works, at the time of nomination, may already incorporate variations to the main works, in which case they will not be treated as variations to the subcontract works. For example, the electrical installation may have been shown on the contract drawings for the main works as having all horizontal conduits in the floor screed. When the nomination is made, the subcontract drawings may show the horizontal conduits in the ceiling space.

This variation (to the main works) may cause considerable reprogramming of all trades in the ceiling space and have an effect on the sequence of partitions and floor screeds. It may be one of the reasons for the subcontractor's programme to be at odds with the contractor's programme. If the variation to the main works is recognised prior to the nomination, and an extension of time is made for it, the contractor may have no need to object to the nominee. If the variation is not recognised prior to the time of nomination, the discrepancy between the contractor's and the subcontractor's programme may have to be resolved between the architect, the contractor and the subcontractor in the light of the variation (after nomination and preferably before the subcontract is made).

If sufficient details were given at tender stage, the type of variation mentioned above ought to be detected by the design team and the contractor. What is the situation if insufficient information is given in the principal contract to enable the contractor to know if the conduits were originally intended to be in the floor or ceiling space? The contractor will have to assume one or the other in order to programme the sequence of trades and to price the work at tender stage. The design team may argue that there is no variation to the main works (particularly if it was always intended that conduits would be in the ceiling space, but this information had not been given to the contractor at tender stage). In most cases the contractor would have a strong case for a variation. The failure to give sufficient information at tender stage may enable contractors to exploit the situation by alleging variations when, in fact, they had made the correct assumptions at tender stage.

Variations to the subcontract works introduced after acceptance of the subcontractor's tender may have cost implications for the subcontractor only, or for the subcontractor and the contractor. Time-related costs may be justified for the subcontractor but not for the contractor. Each variation will need careful analysis by the contractor and the subcontractor in order to ensure that the time and cost effects are detected and notified promptly.

6.10 Delay and disruption claims

Subcontractors are likely to be delayed by various causes. Subcontractor's claims for delay or disruption to the progress of the subcontract works for reasons which give rise to a claim against the employer are likely to receive the contractor's co-operation to ensure that the full effects are reflected in extensions of time and additional payment made under the principal contract. The sooner the contractor and subcontractor can recognise the merits of co-operating on the keeping of records, giving notices and the means of formulating a claim, the greater the chance of maximising the remedy and reimbursement of additional payment. A joint approach which is consistent is a powerful tool, providing that the claim has merit and substance.

However, claims for delay or disruption to the progress of the subcontract works by the contractor, or other subcontractors, are likely to be resisted by the contractor for various reasons:

- If the delay is concurrent with a delay which is the employer's responsibility, the contractor's claim against the employer may be prejudiced;
- The contractor may have difficulty in disentangling the causes and effects of delays caused by himself and/or various other subcontractors, thereby increasing the likelihood that the cost will have to be borne by the contractor.

If the contractor can clearly identify the culprit(s) to whom a subcontractor's claim may directed, he may be less resistant to the claim. Much will depend on the chance of recovering the costs from the defaulting subcontractor(s). Where the contractor is to blame for the delay or disruption, settlement will depend on the contractor's and subcontractor's records and the subcontractor's ability to present his claim with clarity. Onerous subcontract conditions and counterclaims will often feature in negotiations and it may be in the contractor's interest to do a deal in order to conceal the nature of the dispute from the employer's professional advisers (particularly if the delay is one which is concurrent with delays which may give rise to additional payment under the principal contract). Subcontractors who recognise a vulnerable contractor can often achieve a prompt and satisfactory settlement.

7 Response to Claims: Counter-claims

7.1 General policy

No one likes to be on the receiving end of a claim. From the employer's point of view it will mean additional cost by way of loss of revenue and/or additional payments to be made to the contractor. From the point of view of the professional advisers to the employers, it may reflect on the firms' competence in preparing contract documents and on their skills in contracts administration. They may also be faced with additional costs of administration which cannot be recovered from the employer. When contractors receive claims from subcontractors, they will be mindful of the fact that the claim may arise out of their poor organisational skills, in which case they will not be able to obtain reimbursement from the employer or other subcontractors.

Nevertheless, valid claims are a fact of life in modern construction projects. They are an essential feature of small and large contracts and the machinery to deal with them should be regarded as an important element of control. Prompt submission of notices and particulars, followed by a considered response from the recipient as soon as possible will usually facilitate early remedial action and settlement.

The employer's professional advisers will normally be required to act as independent valuer or certifier under the contract and/or advise the employer on the contractor's rights and entitlements. In *Pacific Associates Inc and Another* v. *Baxter and Another* (*supra* – Chapter 1), it was held that the contractor had no recourse against the engineer if he should fail to certify properly and act fairly. The contractor would, however, be able to recover from the employer. Consultants should therefore be aware that they are likely to be the target for negligence claims from the employer if the contractor's claims arise out of their failure to value or certify in accordance with the conditions of contract. Employers should also be aware that their interference with the impartial certifying function of their consultants will be self-defeating (*Morrison-Knudsen* v. *B.C.Hydro & Power* and *Nash Dredging Ltd* v. *Kestrell Marine Ltd,* Chapter 1 – *supra*).

Consultants who fend off claims to avoid criticism of their own performance may only be compounding the problem and laying themselves, and the employer, open to greater claims from contractors. Delay in recognising a claim and responding to it may cause any hope of effective remedial action to be lost. Poor advice given by consultants to the employer upon which the employer relies to embark upon the road to litigation or arbitration which could otherwise have been avoided may lay the consultants open to claims from the employer.

If claims are to be dealt with effectively, employers and their professional team should decide on policy at the outset. There should be a system of referral to experienced staff who are not responsible for the day-to-day administration of the project. Advice from an independent consultant may be appropriate from time to time. A policy statement should include the following:

- Consultation as soon as the first notice from the contractor is received (or as soon as any member of the professional team recognises a potential claim);
- Delegation of responsibilities to verify facts;
- Consultation to determine the validity, merits and substance of the claim;
- Consultation to analyse the causes and effects of the matters which are the subject of the claim;
- Recommendations on the quantum of the claim;
- Content of written response and necessary certificates to be issued.

Whatever policy is adopted, the timing and content of the first response to a claim situation may be critical to its successful conclusion with the minimum exposure to delay and additional cost. It is important that the response should reflect the opinion of the certifier (which may take into account the various matters discussed during consultations with other members of the professional team and the opinions of persons to whom the claim may have been referred).

The content should be sufficiently detailed to show that the matter has been properly considered and the door should be left open to allow the contractor to submit further arguments or facts in support of the claim.

7.2 Extensions of time

Prompt response to any situation which may jeopardise progress and completion of the works by the due date is necessary for practical and contractual reasons. From a practical point of view, it is essential to have a valid programme which is consistent with progress and the latest extended completion date. Without continual review which takes account of actual delay and entitlement to extensions of time, there is no means to plan future issuance of details and instructions and there is no yardstick by which to measure future delays. Extensions of time granted several months after the event (or even several months after completion of the project), are of no practical use and any opportunity which may have existed to reduce the delay may have been lost.

From a contractual point of view, time to exercise the powers to grant an extension may be critical to the employer's rights to levy liquidated damages (*Miller* v. *London County Council*, Chapter 1 – *supra*). Some doubt has been expressed on the validity of the argument that if extensions of time are not granted within the time contemplated by the contract, the employer's rights to liquidated damages are extinguished. In *Temloc Ltd* v. *Erril Properties Ltd*, (Chapter 1 – *supra*), the *employer* argued that since the architect had failed to grant an extension of time within the twelve-week period provided in clause 25.3.3 of JCT80, the employer could not recover liquidated damages but he could recover general damages in lieu of liquidated damages (which in this case had been £nil in the appendix to the contract). The judge took the view that the twelve-week period was *directory only* and not mandatory. This view has been highly criticised by distinguished authors on construction contracts. However, since it was the *employer* who was seeking to rely on this provision in order to recover damages which it could not otherwise claim under the liquidated damages provision in the contract, it is not surprising that the judge did not see fit to allow the employer to benefit from his own architect's failure to grant an extension within the time limits laid down in the contract. If this practice was condoned by the courts, nothing would prevent employers from encouraging architects to delay granting an extension of time if the general damages were found to be greater than the liquidated damages specified in the contract. It is submitted that the contractor would still be able to succeed in arguing that the employer could not rely on the liquidated damages provisions in the contract, if the architect did not grant an extension of time within the twelve-week period, notwithstanding the judge's view in *Temloc* v. *Erril Properties*.

In a recent Australian case, it was held that the employer had the option to levy liquidated damages (if the architect issued the necessary non-completion certificate) or, if no certificate was issued, the employer may levy general damages which may exceed the amount stipulated for liquidated damages, *Baese Pty Ltd* v. *R.A. Bracken Building Pty Ltd* (1989) 52 BLR 130. The commentary to the case (at pp 131 and 132) suggests that the judgement is of limited application and should not be regarded as creating a precedent giving rise to a general right to opt for liquidated damages or general damages.

The requirement to grant an extension of time within the periods contemplated by the contract does not mean that the the architect's, or engineer's opinion must be the right. The architect, or engineer, need only consider the delay and grant, or refuse to grant, an extension of time within the requisite period. Provided that there was a genuine attempt to deal with the matter, and the contractor was notified of the extension, or reasons for refusing an extension, within the period, then the contractual provisions will be satisfied and the employer's rights to rely on the liquidated damages provisions will be preserved. A refusal, or insufficient extension, which is not based on a genuine attempt to assess the delay (but merely to preserve the liquidated damages provisions), may not be effective. No response, or protracted exchanges of correspondence with no conclusion may not preserve the employer's rights to liquidated damages if it should be subsequently held that an extension of time ought to have been granted at the appropriate time.

The contents of a response to a notice or claim for an extension of time are important. Whilst it is not usually necessary to give periods of extension for each separate cause of delay (save to the extent that it may be required separately for a claim for loss and/or expense pursuant to clause 26.3 of JCT80), it is good practice to do so for the following reasons:

- It enables the contractor to be fully aware of the delays which have been considered (within the time limits for granting an extension);
- It facilitates agreement on some of the delays and extensions of time granted therefor, and enables both sides to concentrate on resolving the contentious delays;
- It facilitates agreement on delays which may, in any event, have to be quantified in order to establish the amount of additional payment;
- It enables the contractor to identify which delays apply to which subcontractors so that consistent extensions of time can be granted under each subcontract.

Some common problems which arise are:

Late information

Information which is issued late (having regard to the programme) but does not actually cause delay to the progress of the works because the contractor is not ready to commence the work which is affected by the late information. Is the contractor entitled to an extension of time? Factors to be considered include the following:

- Is there a lead time? That is to say, does the contractor have to order materials or arrange for the work to be done by a sub-contractor? The architect, or engineer, may be already in delay prior to any delay by the contractor and would therefore not have been in a position to anticipate the site progress. It may well be that the information was required before the contractor commenced the affected work and the contractor had no need to commence prior to receiving the information (see Figure 7.1).

- Is the contractor in delay for matters which would justify an extension, or is he being dilatory?

It may be that even if no extension was justified, the employer could not in any event have been in a position to give the information earlier and could not therefore have obtained use of the project any earlier than the time required to complete the remaining work affected by the late information. The best advice is not to rely on the contractor's delays to put off issuance of information for construction. If it is unavoidable, the contractor may be entitled to the benefit of the doubt and the employer may have no claim against the contractor.

Information and variations issued after the completion date

If the contractor is in culpable delay and liable to liquidated damages, further delay caused by information and instructions issued after the completion date has passed may be difficult to deal with within the contractual machinery. In such circumstances, contractors will seize the opportunity to establish extensions of time for the full period up to the date when the delay ceased to affect the progress of the works, plus an allowance to complete the remaining works. Much will depend on the reasons for the late information or variation (see Chapter 6 – *supra*) and the terms of the contract.

If the contract does not provide for extensions of time after the completion date has passed, or if the provisions allow for extensions of time without preservation of the employer's rights to liquidated damages, the employer and his professional advisers will need to give careful consideration to the need for giving any instructions at all, and if they cannot be avoided, what should be done to protect the employer's interests?

If the architect, or engineer, is of the opinion that an extension of time can, and ought to be made, then an extension should be made having regard to the facts and circumstances. If the architect, or engineer, is of the opinion that no extension can be made, then the contractor should be advised accordingly.

Except in the most straightforward of cases, these circumstances may require expert advice on the meaning of the contractual provisions and the period of extension which may be justified.

Omission of work

The provisions of JCT80 contemplate an allowance for any variation, as an omission of work which produces a saving in time, when considering the period of any extension of time which may be granted. Clause 25.3.1.4 requires the architect to state:

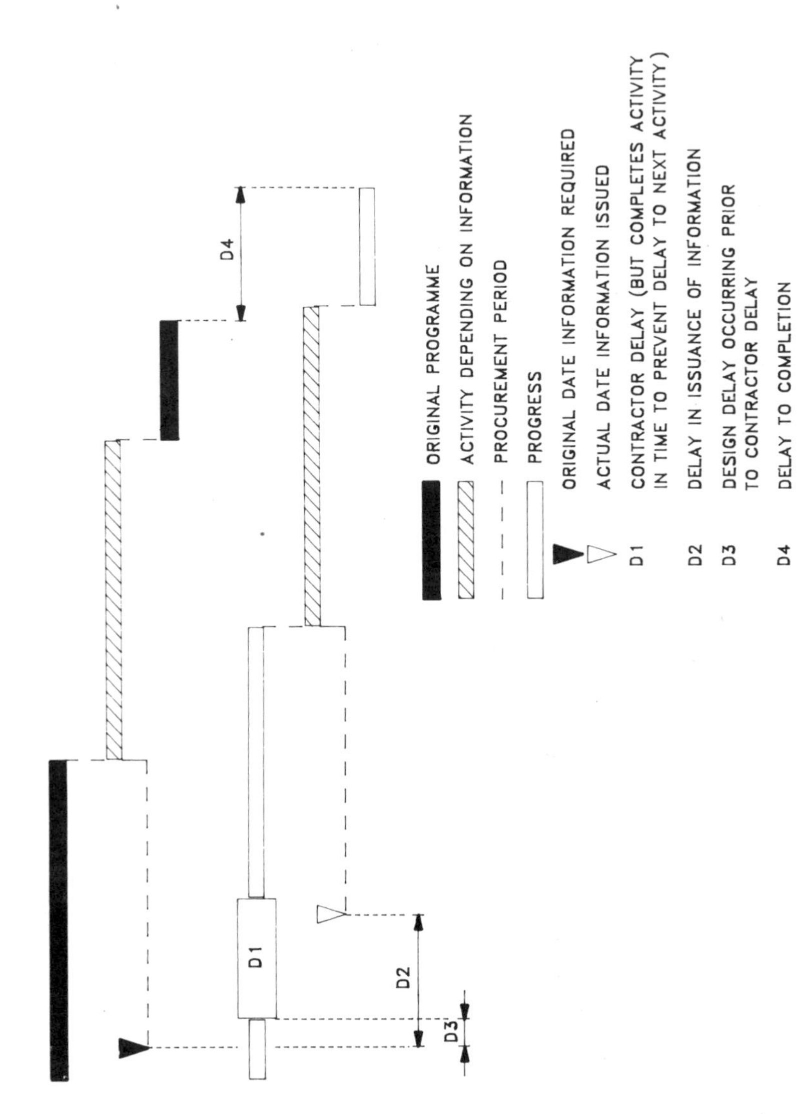

Figure 7.1 Late information concurrent with contractor's delay

'the extent, if any, to which he has had regard to any instruction requiring as a Variation the omission of work *issued since the fixing of the previous Completion Date,*' (emphasis added).

The architect may also, after the completion date, fix an earlier completion date than that previously fixed if it should be reasonable to do so having regard to omissions ordered after the date of fixing the previous completion date – clause 25.3.3.2.

Whether or not there should be any omissions, the architect is required to grant an extension of time within twelve weeks of the contractor's notice, or before the completion date, whichever is earlier. Even if notices and particulars and extensions of time are given without delay, the contractual provisions may not allow all omissions to be taken into account. There may be a period when omissions occur but which cannot be taken into account (see Figure 7.2). While it is reasonable to have provisions to make allowance for omissions, it appears that the JCT80 provisions could be improved to catch other omissions which occur *after the delaying matter which was the subject of the previous extension of time had ceased to operate.*

It should also be borne in mind that, where there is delay in granting an extension of time (even if it should be granted within the requisite period), the contractor may issue a programme which is a fair reflection of the extension due with the exception of any omissions. It would be good policy to bring the omissions to the attention of the contractor before work has progressed in accordance with the revised programme to the extent that the benefit of the omission is lost.

In order to prevent these circumstances arising, where the architect is of the opinion that there is a case to make any allowance for omissions, he should address the matter without delay in consultation with the contractor so that there is no doubt as to the reasonableness of any allowance. In any event, an allowance should only be made where the omission is on the critical path, or is of such a nature that resources (previously required to execute the omitted work) can be diverted to execute work on the critical path *and that there will be a benefit in time.* It is insufficient to make a subjective judgement without a proper analysis of the programme and progress to establish that a saving in time was justified.

It is important to note that omissions to have the work done by others is a breach of contract and may not qualify to be taken into account (see also Chapter 1 – *supra*).

Concurrent delays

Many architects, and engineers, refuse to grant extensions of time for qualifying delays when the contractor is himself in delay at the same time. Sometimes this is justified, but very often an extension of time is necessary (see Chapter 5 – *supra*).

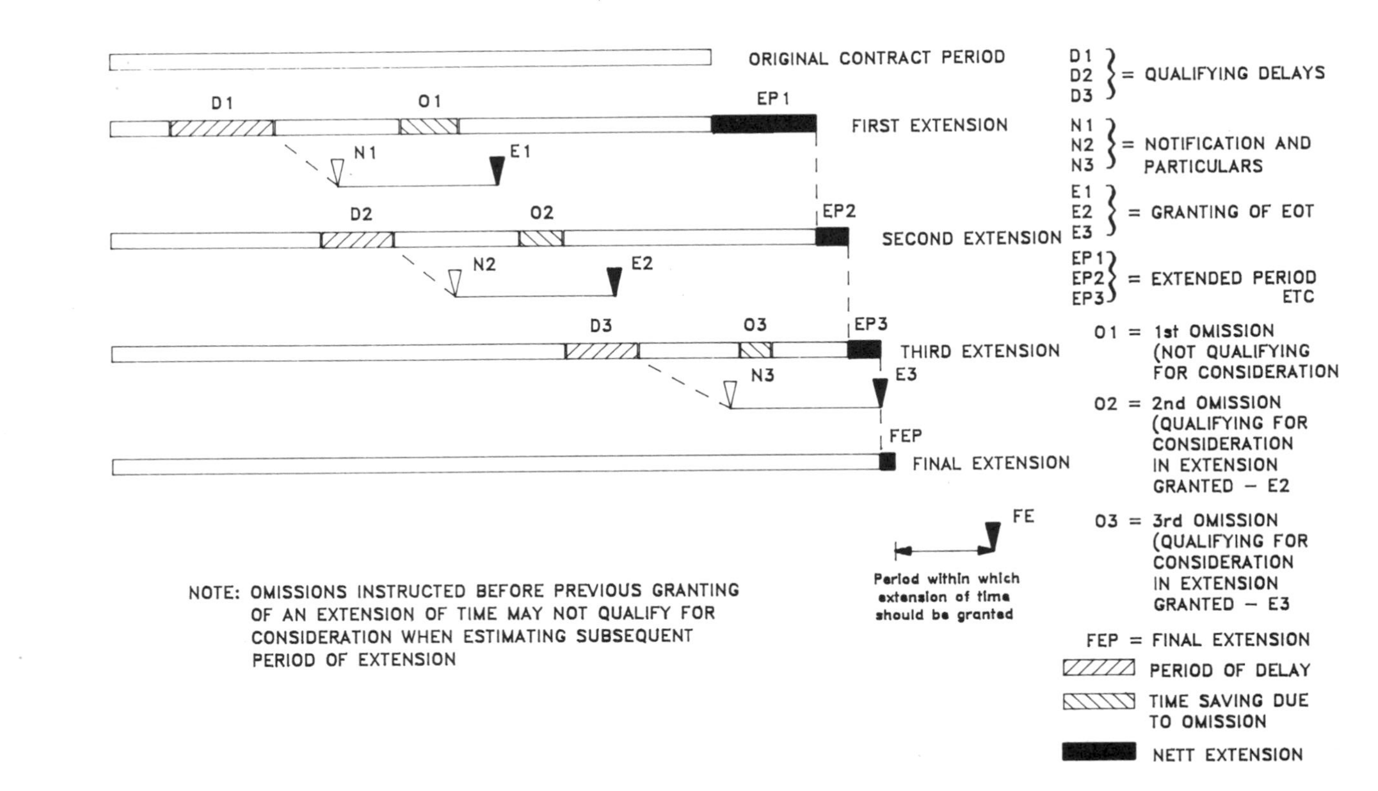

Figure 7.2 Omission of work – clause 25.3.3.2 of JCT80

Once the contractor has given notice of delay, or if the architect, or engineer, is aware of delays on the part of the contractor, it is important that these delays are monitored. The consultants responsible for granting extensions of time and/or certifying additional payment arising out of delay owe a duty of care to the employer to ensure that the contractor is not given any more time or money than is reasonable in all of the circumstances. They will have to consider those matters described in Chapter 5 (*supra*).

In order to ensure that the employer is not exposed to additional costs which should not rightly be borne by the employer, the architect, or engineer, will have to be aware of delays by the contractor at the earliest possible time. Once aware of these delays, it is important to keep contemporary records.

Any response to claims for extensions of time should state which delays (by the contractor) were concurrent with qualifying delays and which (if any) were considered to be delaying completion of the works. This may not necessarily reduce or affect the extension of time to which the contractor is entitled, but the contractor will be aware of the fact that the architect, or engineer, is well informed on the progress of the works.

7.3 Claims for additional payment

While a prompt response to claims for extensions of time is essential for practical reasons, and to keep the liquidated damages provisions alive, a response to claims for additional payment is not usually subject to the same urgency. Nevertheless, provided that the contractor gives notice and particulars in accordance with the contractual provisions, assessment of the sums due and certification for payment should be done as soon as possible. It is often in the employer's interests to deal with these claims as early as possible. Agreement of claims and settlement from time-to-time during the course of the project reduces the contractor's ability to collect all outstanding claims into a 'global claim' which may be little more than a statement claiming the difference between the certified value of all completed work and the actual cost.

Many contractors may prefer to wait until the end of the contract before submitting a formal claim. If that is the case, the employer may not be disposed towards any attempt to encourage the contractor to submit his claims as they arise so that they can be settled and set aside. In such circumstances, the employer's professional team should be aware of potential claims and make whatever assessment they can from their own investigations and records. The employer will be interested in knowing the amount of the potential claim, but no action should be taken to effect payment before the contractor has complied with the contractual procedures (unless a deduction in the contract price may be justified). Once the contractor's particulars are received, the assessment can be modified in the light of such particulars and a prompt settlement may be possible.

If the contractor has gone to a great deal of time and trouble to submit a well thought-out claim, with full particulars and sensible calculations, then a written response merits a similar amount of detail, indicating where there is agreement and reasons for any adjustments which, in the opinion of the architect, or quantity surveyor, or engineer are considered to be appropriate. If, on the other hand, the contractor's submission is poorly argued and presented, the temptation to dismiss the claim out of hand should be resisted. A response should explain why the submission is unsatisfactory and it should give the contractor the opportunity to clarify, or amend the claim. Further particulars may be requested, and these should be specified. If it is a frivolous, or unfounded claim, the contractor should be politely told so. If the claim is justified, and has merit, it is unlikely to go away, in which case it may be appropriate to give the contractor some guidance as to presentation. It may well be that the matter which is the subject of the contractor's claim is one which ought to be dealt with as a variation, thereby giving the engineer, or quantity surveyor, the scope to deal with the matter within the rules for valuation of variations. Provided that the employer is not disadvantaged, this approach may be the most acceptable to all concerned.

7.4 Counter-claims: liquidated damages: general damages

Many claims which may be levied by the employer against contractors are overlooked or are not considered to be worth pursuing. This may be because employers are fearful that such claims could be the reason for large claims by contractors which may otherwise have been waived.

Claims which may be levied against contractors include those arising out of defective work and failure by the contractor to execute work expressly authorised under the terms of the contract. Some claims may be made under the terms of the contract and the amounts of the claims may be set off against interim or final payments due to the contractor from the employer. Others may be common law claims.

The most common counter-claim against contractors is the deduction of liquidated damages for late completion of the works (or if provided for in the contract, for late completion of sections of the works). In order to be enforceable, a liquidated damages provision must be unambiguous and the sum stated in the contract must be a genuine pre-estimate of the employer's likely loss, estimated at the time of making the contract in the event of delay to completion. If the sum stated is a penalty, the employer cannot rely on the clause. It will not be deemed to be a penalty merely because the employer's actual loss is less than the liquidated damages (for example, if the liquidated damages were based on realistic anticipated rents at the time of making the contract, and the market had collapsed by the time the works were complete, the contractor could not argue that the sum was a penalty).

The employer's professional team may have to advise the employer on the amount of liquidated damages to be inserted in the contract and on the contractor's potential liability for liquidated damages when the contractor is in delay during the course of the contract. However, consultants should not use the threat of liquidated damages in any response to a contractor's delay claim, even if it is clear that the contractor is in default. Such matters should be for the employer alone, and then only when the consultants have properly considered all delays which may give rise to an extension of time.

JCT63 required the architect to issue a certificate stating that in his opinion the works ought reasonably have been completed by the date for completion as a precondition to the employer's rights to deduct liquidated damages – clause 22. Having regard to circumstances which may have arisen during the course of the contract (such as delay by the employer which may not have qualified for an extension of time) the architect may have had good reason not to be able to express such an opinion, in which case no certificate could be issued and no liquidated damages could be deducted. JCT80 only requires the architect to certify that the contractor had failed to complete the works by the completion date (as a fact) before the employer can deduct liquidated damages – clause 24. Many other forms of contract do not require a certificate of any sort as a prerequisite to the employer exercising its rights to deduct liquidated damages.

It is often argued that the architect cannot certify that the contractor has failed to complete the works by the completion date unless and until he has considered all of the delays for which an extension of time may be granted, *Token Construction Co Ltd* v. *Charlton Estates Ltd* (1976) 1 BLR 48. If, however, a further extension of time is granted after liquidated damages have been deducted, the employer must repay the liquidated damages for the relevant period of further extension (for example, clause 24.2.2 of JCT80). The contractor is entitled to interest on the liquidated damages withheld, and subsequently repaid, *Department of Environment for Northern Ireland* v. *Farrans* (1981) 19 BLR 1. Clause 47(5) of the sixth edition of the ICE conditions of contract provides for interest on liquidated damages to be repaid to the contractor as a result of further extensions of time.

If there are no provisions in the contract for liquidated damages the employer may be able to levy a claim for general damages. Where there is a provision for liquidated damages for late completion of the works, but there are no provisions to deduct liquidated damages for late completion of each phase (assuming that the contract contemplates phased completion), the employer may have a claim for general damages for late completion of any phase, *Mathind Ltd* v. *E. Turner & Sons Ltd*, (see Chapter 3 – *supra*). Where the employer has lost his rights to liquidated damages, he may be able to claim general damages for late completion (see Chapter 1 – *supra*).

General damages may arise if the employer suffers loss as a result of any breach of contract by the contractor. Provided that the nature and cause of the loss are not identical to those which may be recovered under a liquidated damages provision, then general damages may be recoverable in addition to the liquidated damages for late completion. Some tailor-made conditions of contract provide for liquidated damages *and* general damages for delay. Provided that the nature of the damages are not identical (thereby duplicating the claim for delay), provisions of this kind may be enforceable. For example, if the liquidated damages were a genuine pre-estimate of the loss of revenue and direct costs of supervision during the period of overrun, a separate claim to recover delay costs levied by other contractors (who were delayed by the contractor) would not be a duplication of the same damages and may be recoverable.

7.5 Claims against subcontractors

There is an increasing incidence of claims made by subcontractors against contractors and by contractors against subcontractors. Some forms of subcontract devised by contractors are aimed at precluding any claim at all from subcontractors and they attempt to provide for claims to be made against subcontractors on dubious grounds with little supporting evidence. Recent cases in the courts have identified the most unreasonable contractors in this regard. Notwithstanding the adverse publicity and understandable indignation expressed by various trade associations, the majority of contractors use recognised standard forms of subcontract and apply the provisions fairly.

Where a subcontractor is in delay, or is disrupting the progress of the works, the contractor will naturally wish to recover any losses incurred from the defaulting subcontractor. Where there is only one subcontractor in delay, and there are no competing delays, it is possible to establish liability with relative ease. However, it is probable that there will be several delays occurring at the same time, in which case the contractor will be faced with the difficulties which have been mentioned in respect of concurrent delays in Chapter 5 (*supra*). Only the most careful attention to records and regular updating of programme and progress schedules will enable the contractor to establish liability and quantum of damages which may be recoverable from several subcontractors (and possibly from the employer) for what may be substantially the same period of delay.

Where the contractor becomes liable to liquidated damages for late completion of the main works, he will seek to recover some, or all of the damages from defaulting subcontractors. In the case of nominated subcontractors, this may not arise (for example, where the contractor is able to obtain an extension of time for delay on the part of nominated subcontractors). Nevertheless, the contractor may have a claim against the nominated subcontractor for the costs of prolongation which he could not recover from the employer.

Apportionment in the event of delay by several subcontractors is almost bound to cause difficulty. Even where the contractor has been able to calculate the sum which is due from the subcontractor, the provisions for set-off in the subcontract may frustrate the contractor's ability to deduct the amounts due from payments which would otherwise be paid to the subcontractor. The general rule is that the contractor's rights to set-off at common law are not affected by the contractual provisions unless there is clear language in the contract to bar the general right of set-off, *Gilbert Ash (Northern) Ltd* v. *Modern Engineering (Bristol) Ltd* [1974] AC 689. However, where the terms are explicit, and the set-off provisions are exclusively laid down in the subcontract, the contractor's rights to set-off will be determined by the contractual provisions.

An architect's certificate of delay or non-completion by a nominated subcontractor may be a prerequisite to the contractor's rights to damages from the defaulting subcontractor under JCT 80. This can be troublesome, particularly where the architect refuses to give permission to grant an extension of time (for any reason) to a subcontractor and at the same time will not issue a certificate of non-completion against the subcontractor, *Hong Kong Teakwood Limited* v. *Shui On Construction Company Limited* (1984) HKLR 235.

The *Shui On* case was, however, rather different from most situations found in the United Kingdom. In the first place, the provision in the Hong Kong equivalent of JCT63 to permit extensions of time for delay on the part of a nominated subcontractor had been deleted and, in the second place, the subcontract between *Shui On* and *Hong Kong Teakwood* contained a 'pay when paid' clause. An almost identical situation arose in *Schindler Lifts (H.K) Ltd* v. *Shui On Construction Company Limited* (1984) 29 BLR 95. Here, the architect issued a certificate of non-completion against the contractor, but not against the subcontractor. The employer deducted liquidated damages from the payment certificates issued in favour of the contractor after the certificate of non-completion. The payment certificates included sums in favour of the subcontractor. The contractor argued that he had not received payment from the employer, and since the obligation to pay the subcontractor did not arise until such time as payment was received from the employer, no payment was due to be made to the subcontractor. The Court of Appeal in Hong Kong found in favour of the contractor. This did not mean that the subcontractor had no remedy. There were provisions for arbitration in the principal contract and in the subcontract and the disputes between the parties were capable of resolution in arbitration.

In addition to claiming all, or part, of the liquidated damages for late completion of the main works from a defaulting subcontractor, the contractor may also have a claim for other loss and expense, such as prolongation and/or disruption costs incurred by the contractor and by other subcontractors. The quantification of such claims where there are several competing delays is bound to be fraught with problems and unless a commercial settlement can be reached between the contractor and the subcontractors, the matter may have to be settled by several separate arbitrations or by the same proceedings involving several parties.

8 Avoidance, Resolution and Settlement of Disputes

8.1 Commercial attitude and policy

Many contractors and subcontractors genuinely wish to avoid claims even when there are good grounds for them. This attitude is usually adopted in the belief that firms with a reputation for claims will not be included on some tender lists, and where they are included, they may be disadvantaged if tenders are very close. In some sectors of the industry, firms may be justified in believing that a history of claims will be a dominant feature in the evaluation of their suitability for new projects. However, provided that the firm submitting the claim follows some simple rules, there is no reason to suppose that the pursuit of valid claims is detrimental in the long term.

It is, of course, very helpful if the contractor has done a good job, finishing as soon as was reasonably possible, and has co-operated with the employer and the design team. However, if the contractor has submitted a poor tender, underestimated the complexity and/or underresourced the project, his claim may well be seen by the recipient as a means to recover some of the contractor's losses caused by a poor tender and poor management. It is quite natural, in these circumstances, for the employer and his professional advisers to suspect the contractor of employing a pricing policy to obtain work with the intention of using every possible means to recover a much larger sum when the project is complete. It is not surprising if relations between the parties deteriorate almost before the ink on the first interim payment certificate has dried. Very often, this policy will be obvious to the design team if the contractor is complaining of late information at every opportunity even when it is clear that no delay will be caused. Every letter will be an attempt to create evidence for a dubious claim at some future date.

On the other hand, a contractor with a valid claim will be doing himself no favours if he proceeds reasonably well with the project and co-operates with the employer and consultants, but hardly mentions the fact that he intends to submit a claim until the end of the job (usually after he has been able to persuade the architect, or engineer, to grant a reasonable extension of time based on inadequate notices and particulars). Some contractors adopt this policy purely to maintain good relations or in the hope that a favourable opinion on extensions of time and/or borderline compliance with specifications will be forthcoming. It may be expecting too much to believe that the consultant will form a favourable opinion about a substantial claim for additional payment when the consultant has not been given any information to enable the employer to make provision for payment.

The contractor who does a good job and properly manages the project, will often stimulate the design team to perform well. If, at the same time, the contractor gives notices and particulars in accordance with the contract, avoiding provocative language and frivolous claims, then he is more likely to be able to resolve his claims painlessly.

Even when contractors have, for commercial reasons, made a policy decision not to submit a valid claim, this policy will be soon be reversed if the employer decides to levy a claim for liquidated damages after an insufficient extension of time has been granted. Many consultants and employers have underestimated the potential for the contractor to claim considerable sums of money when he is forced into a corner. For this reason, the employer's professional advisers should monitor all potential claims for extensions of time and additional payment, so that the employer can consider the risks and advantages of levying a claim for liquidated damages. It may be a better decision not to levy a valid claim for liquidated damages if the potential claim from the contractor will far outweigh the claim for liquidated damages. If the contract contains provisions to bar the contractor's claims (failure to give notice and the like), the employer's decision to levy liquidated damages may not be influenced in the same way.

8.2 Claim submissions

Unfortunately, the evaluation of claims is not an exact science. The basis of calculation is dependent on a complex interaction of factors which may be unique to the project. The contractor's method of pricing, allocation of prime cost and overheads in the tender and in the accounting practice, programme, methods of construction, records, monitoring and control systems all have a part to play in the evaluation process. If the contractor has an integrated computerised costing and accounting system with a sensible allocation of cost codes, the evaluation process may be simplified. If the accounting system comprises too many categories it may suffer from a higher incidence of wrongly allocated costs. On the other hand, too few categories may be of no use, thereby necessitating the laborious task of searching through all of the source documents.

Whatever the standard of records and management accounts, even if it is possible to calculate, with precision, the correct amount of the claim, it is a fact of life that the claim is unlikely to be paid in full. For this reason, even the most professionally prepared claim will include a measure of over-valuation as a negotiating margin.

If the contractor has complied with all of the contractual provisions for claims, the employer's professional advisers may be well advised to settle them during the course of the project, leaving very little to resolve at the end. If this cannot be done, the final claim will probably contain a large negotiating element.

The first submission of a claim requires very careful planning. It must not contain any information, assumptions or calculations which can be used against the party submitting the claim. Several alternative approaches may be necessary in order to establish which is the best and most persuasive presentation. It is important to carry out several crosschecks to ensure that the financial data and assumptions can stand up to scrutiny by the recipient. Whilst there may be justifiable reasons for actual prolongation costs to far exceed those which may have been possible to derive from the rates for preliminaries in the bills of quantities, it is often an uphill battle to persuade the recipient that the additional costs are a direct result of matters for which the employer is responsible. The contractor may be well advised to anticipate the steps which may be taken by the opposition when scrutinising the claim. Reliance upon the recipient's inexperience and lack of knowledge in the hope of gaining an advantage may be self-defeating. If there is an element in the claim which is found to be dishonest, then the remainder of the claim, no matter how well founded, is likely to be treated with the extra caution which it deserves.

How then, is the contractor to include sufficient margin in his claim to allow for negotiation and at the same time avoid criticism for appearing to be disreputable? Should he include elements which are fairly obvious candidates for rejection so that they can appear to be the basis of the first compromise, leaving the way open for some of the 'grey areas' to be argued vigorously? It is not unusual for some very dubious elements of a claim to succeed merely because they are more palatable to the recipient than other elements which may reflect on the performance of the design team (and which are rejected).

In spite of the fact that a reputable contractor, or his appointed claims adviser, will not deliberately wish to submit a claim which contains dubious elements, they will be aware that it is necessary to include substantial sums in the claim which are expected to be rejected at some stage of negotiations. In some cases, not all of the dubious elements will be rejected, in which case the contractor will recover more than that to which he is entitled. In the long term, the contractor may not be any better off because many claims will be settled below a sum which reflects his full entitlement. Unfortunately, some employers will benefit at the expense of others.

The person, or persons, responsible for preparing the claim will have to establish the basis and quantum of claim which is considered to be correct in all respects. This will take into account all of the facts and particulars which are available and reasonable assumptions where they are necessary.

The lowest and highest sums which are likely to be awarded if the matter should proceed to arbitration should be considered, giving each head of claim a rating in order of merit. In cases where there is no evidence of concurrent delay and the contractor has excellent records, it may be possible to quantify prolongation costs with a high degree of certainty. If this is the case, the likely success factor of this head of claim may be as high as one hundred per cent. If there is concurrent delay and incomplete records, the success factor of this head of claim will be reduced accordingly. Claims for disruption will rarely justify a one hundred per cent chance of success.

However, such claims which are based on a logical analysis, where cause and effect are established, will be at the high end of the probability scale. Claims which tend to be based on a global assessment will normally be at the lower end of the probability scale. That is not to say that global claims, in the appropriate circumstances, will not merit a high rating. Some claims for finance charges will be well founded in contract, or in law, whilst others may be less likely to succeed. The likelihood of recovering the cost of preparing the claim may be zero. In some cases this head of claim may be justified, even if the probability of success is unpredictable.

Having established the likely range of success of the 'real' claim, it will be necessary to decide how, and to what extent, the negotiating margin can be added. This is not an easy task. If experience has shown that some settlements fall below fifty per cent of the original claim, the contractor is faced with finding plausible methods to to double the amount of his first submission. The idealist will view this process with some distaste. The commercial realist will know that it is unavoidable and all of his experience and imagination will be called upon to ensure that the negotiating margin is at least arguable.

Every 'grey area' must be presented as black, or white, depending on the circumstances. Care should be taken to avoid presenting black as white. Under no circumstances should contemporary records be changed, or invented, in order to distort the truth. Dishonesty should be avoided at all costs. The contractor, or subcontractor, submitting the claim should be aware of the probable range of success, the nature and quantum of the negotiating margin and the strengths and weaknesses of the claim before submission. Any elements which cannot be argued with at least some degree of conviction may have to be discarded.

Most contractors, and subcontractors, will wish to reach an amicable settlement. Some will have decided, before submission of the claim, that under no circumstances will they take the matter to arbitration if settlement cannot be reached. This attitude is often brought about by the high cost of arbitration, particularly if previous experience has shown that the unrecovered costs of arbitration have not been justified in the light of the award. If this attitude exists, then the negotiating margin is likely to be higher than that which may otherwise have been added. It is, of course, fatal to let the opposition discover that arbitration has been ruled out. If the case is sound, the contractor may be persuaded to contemplate arbitration at the outset (if the matter cannot be settled). In these circumstances, the negotiating margin may not be excessive. If there are a number of substantial 'grey areas' in the claim, some employers (particularly government bodies) may have no option but to arbitrate, even if there is a willingness to settle. This must be taken into account at the outset.

Many contractors have the resources and capability to prepare their own claims. However, even the best organised contractors (including those who are recognised as being amongst the leading companies in the industry) are often unable to make the most of their case in a written submission. Whilst a poor claim cannot be made into a good one, a good claim can easily fail if it is presented badly. Many good claims fail, at least in part, because the author of the claim is influenced by staff in the company who have vested interests in overlooking any shortcomings in the contractor's case and perhaps by placing too much emphasis on elements of the claim which have caused dispute throughout the contract. If the contractor's staff have been advising management that the claim is well founded and worth several hundred thousand pounds, they will be reluctant to change their view even in the light of valid counter arguments put forward by the other side.

Many final submissions repeat what has already been said, and rejected, in numerous exchanges of correspondence over several months. Even if the contractor is right, it is important to search for alternative arguments and means of persuasion. This is usually difficult to achieve by staff who have lived with the project and have fixed ideas on what happened and who was to blame. In any event, it is good practice to get an independent view of the strengths and weaknesses of the claim, the likely range of settlement, or award, and expert advice on how it should be presented before any submission is finalised for dispatch to the opposition. If there is any potential liability for liquidated or general damages, this should be brought to the attention of management and taken into account in the overall assessment of the likely recovery.

Once the claim is submitted, the contractor will need to ensure that there is a response or some other means of moving forward. The covering letter to the submission should summarise the claim so that any person who is not familiar with the detail, and who may be making important decisions, can appreciate the nature and amount of the claim without reading the detailed submission and appendices. The letter should invite a reply within a reasonable specified period. It may be useful to suggest a meeting to discuss and explain the claim in more detail before a formal reply is expected.

8.3 Negotiation

If the contractor has a valid case, given notices in accordance with the contract, kept accurate contemporary records and presented his case in a logical and professional manner, he will be starting from a position of strength. If a valid claim is not accompanied by these essential ingredients, the recipient will have little difficulty in finding reasons to reject it.

Whatever the merits of the claim, the initial response will usually concede very little. The contents of the response may be positive, giving cause for optimism, or it may be totally negative, rejecting every aspect of the claim. The former will enable both sides to move forward, whilst the latter will form a barrier to any early progress to resolve the matter. If there is no response at all, or if a negative response cannot be countered by some means of opening a dialogue, the contractor may have little option but to commence proceedings. If he has not already already obtained advice before submitting the claim, the contractor should obtain the advice of experts before taking a decision to initiate formal proceedings.

If the response is positive and negotiations commence, then both parties may be able to settle the matters reasonably quickly. The contractor must be wary of employers who are merely going through the motions with no intention to settle at a reasonable figure. Their tactics will be to find out what concessions are on the table and to waste time. A delayed settlement usually means less in real terms, irrespective of any financing element which may ultimately be included (if any). If there are reasonable grounds to suspect that the employer is not genuinely seeking a fair settlement, the decision to commence formal proceedings should be taken sooner rather than later.

Negotiations may be conducted on an open basis (that is to say that the records of the negotiations may be used by the parties in any proceedings), or they may be *without prejudice* (that is to say that they cannot be referred to in any proceedings). In most cases, *without prejudice* negotiations are more satisfactory as they enable the parties to be more frank and they facilitate concessions which can be withdrawn if the other party refuses to make any concession. If there is agreement on any section of the claim, the contractor should endeavour to persuade the employer to make the agreement open and certify any sums which ought to flow from it. The employer will usually resist on the grounds that he will require an overall settlement.

From the employer's point of view, he will be prepared for the contractor's claim if he has been informed by his professional team pursuant to the contractor's previous notices. Even if the contractor has not complied in all respects with the contract to notify the employer's architect, or engineer, the employer ought to have been made aware of potential claims by his consultants. If he is properly advised, he will already have an outline defence to many of the contractor's claims. If the contractual provisions have been followed to the letter, any sums which are, in the opinion of the architect, or engineer, due to the contractor, will have been certified and paid. In practice, in spite of the problems caused by interference by the employer, the architect, or engineer, may be unable to act freely. This is sometimes the case where the architect, or engineer, is an employee of the employer.

Whoever represents the parties at negotiations, it is important to establish at the outset if they have the authority to make an agreement. Negotiations between staff who are not authorised to finalise an agreement may be suitable for initial discussions, but serious negotiations to conclude a settlement must be conducted by staff with full authority to agree on all aspects of the claim. It is particularly important for the contractor to establish whether, or not, the employer's consultants have such authority (they will not normally have this authority as part of their usual agreement with the employer to provide professional services).

If the consultant has such authority, it should be remembered that he stands to be shot at from both sides. If he wrongly certifies, or negotiates a settlement, to the detriment of the employer, he may be sued for negligence by the employer. If he wrongly certifies to the detriment of the contractor, or fails to negotiate a settlement which is satisfactory to the contractor, he may be exposing the employer to unnecessary costs of arbitration or litigation. Finding the right solution may require a careful and critical review of the consultant's own conduct during the contract and possibly acknowledging mistakes which have been made from time-to-time. For this reason, the employer may be well advised to be represented by an experienced negotiator who has not been involved with the day-to-day administration of the project and who is not tied by previous decisions.

Both parties should decide on the team which will be present to advise and support the negotiator. The temptation to field a large team should be resisted. It is important to select a team who are fully conversant with the matters under negotiation. It should be possible to verify or reject allegations, facts, matters of law or contract, principles of evaluation and the like by reference to members of the team. The negotiator should decide whether any difficult points should be discussed in the presence of the other party, or if negotiations should adjourn to enable private discussions to take place. The team should not be changed unless there is a clash of personalities which is hindering a settlement. Sufficient time should be allowed to prepare for each meeting and a common approach should be established so that no divisions between team members will become evident at the meetings.

Concessions should be considered before any meeting and the negotiator should be ready to concede at the appropriate time if it should be necessary to do so. Concessions should not be made too lightly, and then only if the other party is showing a willingness to give ground. It may not be the best policy to concede too many points unconditionally. At the end of negotiations, both parties will seek a satisfactory overall settlement. Too much given away on individual heads of claim may make it impossible to agree on the entire claim.

If one of the parties is not genuinely seeking a fair settlement, they may field a team which does not have authority and who have to report to others to verify facts or decide on important points. Perhaps they will change the negotiatiing team when it seemed that progress was being made and the other party finds that the entire process has to begin from first base again. If this becomes evident, it may be appropriate to break off negotiations and commence proceedings without delay. It may be worthwhile preparing a notice of arbitration to issue at the meeting, perhaps leaving the door open to serious negotiations in a parting statement to the team leader.

If an agreement can be reached, it is important to have the terms of the agreement recorded and signed by the authorised representatives of the parties before the meeting is concluded. The agreement should make it clear that it covers all matters which were the subject of the negotiations, and if both parties intended the agreement to cover every claim and counterclaim (so that no other claims could be brought against the parties) it should clearly say so. Indemnities may be required with respect to possible claims from subcontractors and/or third parties. The date of payment should be specified and there should be provision for interest to be added in the event of late payment.

8.4 Resolution of disputes by third parties

If, despite all efforts to come to an amicable agreement, no agreement can be reached, it may be possible to resolve the dispute by an independent third party. The most common means of resolution are litigation or arbitration. However, in view of the high costs of litigation and arbitration (the latter now often being conducted with all the formality of court proceedings), alternative means of resolving disputes are becoming increasingly popular. Possible methods include:

Third party expert opinion

One of the parties (usually the employer, if he is serious about settlement) will engage a third party expert to assess the merits and quantum of the claim. If this process is to succeed in facilitating a move to settle, it is important that the expert is given a free hand to come to an impartial view, even if it means criticism of the party who engaged him. After the expert's initial assessment, he may be asked to give opinion on the range within which an arbitrator would probably make an award and on the likely costs of arbitration. This information is invaluable as a basis for further negotiations which may be conducted on a *'without prejudice'* basis. If negotiations fail, and provided that the expert's independent view is likely to be helpful to the case, he may continue and appear as an expert in arbitration or any other proceedings which take place to resolve the dispute.

Conciliation

If the parties are really willing to settle, but there are genuine obstacles to settlement, it may be possible to close the gap between the parties and facilitate a settlement by the process of conciliation. This method may not be imposed unilaterally and the agreement of the parties is essential. It involves the appointment of an independent third party, mutually agreed by the parties, to hear both parties' points of view. The conciliator will usually be a recognised expert on the matters in dispute and he will look at the evidence and listen to the arguments put forward by each side. He will contribute his own ideas on the merits of the case. He will not meet any party in private and all discussions take place with both parties present. The parties may have legal advisers present at any meetings, and they may, of course, meet each other without the conciliator being present. The conciliator's aim will be to bring the two sides together to discuss all aspects of the matters in dispute and lead them to an amicable settlement. The conciliator will not make decisions, but he may make recommendations. It is up to the parties to agree on an acceptable settlement. They are not obliged to agree, and if settlement cannot be reached, the parties may pursue the matter in arbitration or litigation.

Mediation

This process is similar to conciliation. However, the mediator normally meets the parties separately and he may be empowered, if the parties cannot be persuaded to agree, to make a recommendation on the matters in dispute. Any confidential information which is made available to the mediator at private meetings with one party cannot be divulged to the other party. While not usually being conducted in the formal manner normally associated with arbitration, mediation proceedings may be conducted with lawyers and other experts to present each parties' case to the mediator. The mediator will endeavour to find common ground at these separate meetings and he will try to find means of reaching a settlement. A meeting with both parties present will usually be required at some stage. Whoever represents the parties at these discussions, it is essential that they have the authority to agree and settle the dispute. Failing agreement, the mediator may decide on the matters in dispute. The parties are not normally bound by the mediator's decision. However, there is no impediment to the parties agreeing, at the outset of these proceedings, to accept the mediator's decision as final and binding. It is important to consider the nature of the dispute before agreeing that the mediator's decision is to be final. Disputes which involve quantum only may be suitable, whereas disputes which may turn on legal issues would not normally be suitable without a right of appeal.

There are several other methods of resolving disputes, some of which are variations to the above examples, and some of which are almost akin to arbitration. Some contracts expressly provide for disputes to be dealt with by an alternative method, for example the ICE conditions of contract, sixth edition – clause 66(5). Any third party appointed to resolve the dispute by one of these methods is not eligible for appointment as arbitrator in any subsequent proceedings.

Arbitration

Arbitration in England is governed by the Arbitration Acts of 1950, 1975 and 1979. Different provisions apply in Scotland where arbitrations are governed by the (Scotland) Arbitration Act of 1894. The main differences are with respect to questions of law and the enforcement of the arbitration agreement. In Scotland the arbiter has wide powers to decide questions of law and a stay of proceedings is mandatory in Scotland.

The parties' agreement is essential before any dispute can be settled by arbitration. Agreement can be made at any time, but it is usual practice for the agreement to be made at the time of entering into the contract for the work. Standard forms of contract have express provisions for arbitration in the articles or in the conditions of contract.

In the event of there being valid arbitration provisions in the contract which cover the matters in dispute, the parties will generally be prevented from having the dispute resolved by litigation. However, if one of the parties commences litigation, and the other party does not, before taking any steps in the litigation, apply to the courts for a stay of proceedings under Section 4 of the Arbitration Act of 1950, which provides:

> 'If any party to an arbitration agreement, or any person claiming through or under him, commences any legal proceedings in any court against any other party to the agreement, or any person claiming through or under him, in respect of any matter agreed to be referred, any party to those legal proceedings may at any time after the appearance, and before delivering any pleadings or taking any steps in the proceedings, apply to that court to stay the proceedings, and that court or a judge thereof, if satisfied that there is no sufficient reason why the matter should not be referred in accordance with the agreement, and that the applicant was, at the time when proceedings commenced, and still remains, ready and willing to do all things necessary to the proper conduct of the arbitration, may make an order of staying the proceedings.'

then the dispute may be settled by litigation.

If, before taking any steps in the litigation, an application to stay the proceedings is made, then providing that the applicant is ready and willing to have the dispute settled by arbitration, the power to order a stay of proceedings is usually exercised. A stay of proceedings may be refused for the following reasons:

- The arbitration agreement does not contain provisions for immediate arbitration;
- The matters in dispute do not fall within the ambit of the arbitration agreement;
- There would be undue hardship on the plaintiff if the stay were granted:
- The only matter to be decided in the dispute was a question of law;
- Fraud is alleged;
- If there would be two separate sets of proceedings requiring resolution based upon the same facts, one of which would be settled in the courts, and the dispute which was the subject of the application for a stay (if no stay were granted) would be settled in arbitration.

Care should be taken when deciding to avoid arbitration and to proceed in the courts. In most cases the courts do not have the same powers as an arbitrator and they cannot open up, or review, an architect's certificate, *North West Regional Health Authority* v. *Derek Crouch* [1984] 2 WLR 676. Some forms of contract do not restrict the power of the courts. The Singapore Institute of Architect's form of contract expressly states that the courts shall have the same powers as an arbitrator – clause 37(4). The *Courts and Legal Services Act 1990* provides that the High Court may, if all parties agree, exercise the same powers as those conferred upon an arbitrator (section 100, giving effect to an additional section 43A in *the Supreme Court Act 1981*). Other important matters to be considered are the facts that arbitration is held in private and the costs are likely to (but not necessarily) be less than litigation.

When one of the parties has decided to refer a dispute to arbitration, the most important decision is to select the most appropriate arbitrator. If the resolution of the dispute is likely to turn on questions of law, a legally qualified arbitrator may be the best choice. The parties may appoint a judge (in a private capacity) or a circuit judge of the Commercial Court in the Queen's Bench Division of the High Court, pursuant to the Administration of Justice Act of 1970. If the dispute is mainly to do with technical matters, then a technical arbitrator may be more appropriate. If the parties agree, a legal assessor, or a technical assessor can be appointed to facilitate resolution of the dispute. However, the arbitrator must make his own decision, whatever the advice given by the assessor.

If the parties cannot agree on the arbitrator, there is provision in most standard forms of contract for an appointing body (stipulated in the contract) to appoint an arbitrator. Failure to agree on an arbitrator is usually caused by the respondent's desire to delay the proceedings. The disadvantage of having a arbitrator appointed by a third party is that the appointed arbitrator may be a person which neither party would have selected. There may, of course, be valid reasons to object to the other party's choice of arbitrator:

- There may be a conflict of interest (this would in any event be brought to the attention of the parties by the arbitrator);
- The arbitrator may have a reputation for deciding the matters in dispute which is against the interests of the objecting party (in some cases, the arbitrator's views are well known from published works);
- The arbitrator may have a reputation for poor control of arbitration proceedings, thereby permitting delays to occur and costs to increase (a reluctant party may prefer such an arbitrator).

Some forms of contract specify the procedure to be used in the arbitration. The most common procedures in use in the construction industry are the ICE Arbitration Procedure (1983) and the JCT Arbitration Rules. The former is mandatory pursuant to clause 66(8) of the sixth edition of the ICE conditions of contract and the latter is mandatory pursuant to clause 41.9 of JCT80.

In the absence of a specified procedure in the contract, the arbitration will probably include the following stages:

Preliminary meeting

This will formalise the appointment of the arbitrator and a preliminary timetable will usually be drawn up. If the parties can agree a timetable in advance, this will save time and cost of the meeting;

Pleadings

These set out the matters in dispute, the facts and the contractual and legal provisions relied upon. The sequence is as follows:

- Claimant submits points of claim;
- Respondent submits points of defence and counter-claim (if any);
- Claimant submits points of reply to the defence and defence to counter-claim;
- Respondent submits points of reply to defence to counter-claim.

Discovery of documents

After close of pleadings, each party is required to prepare lists of documents for inspection by the other party. In most disputes, discovery may be limited to documents which are relevant to the issues in dispute. In some cases, all documents may have to be disclosed (*general discovery*). Documents which must be disclosed include those relied upon by the parties and any other documents which may be detrimental to the case, or of assistance to the other party's case. There is a strict duty to disclose any and all material no matter how much it may be against the interests of the party having possession, power, or control over the documents. Privileged documents (*without prejudice* correspondence and certain documents which pass between the parties and their legal advisers) should also be listed, but they should not be made available for inspection by the other party.

Inspection of the other party's documents is an important process, and should be done by someone who is experienced and knowledgeable about the matters in dispute. It is equally important to look for anything which is missing, but which should exist. A list of documents which are required should be made and a request for copies should be sent to the other party.

Agreed bundles

After collecting all of the relevant documents, those documents which will be referred to in the hearing are collected and filed in a logical sequence in several bundles. Normally the claimant will prepare the bundles, and the respondent will be given the opportunity to add further documents. The completed files are known as 'agreed bundles'.

Witnesses: proofs of evidence

Witnesses of fact will have an important part to play, particularly if there are gaps in the written evidence. It is important that such witnesses should be selected for their first hand knowledge of the matters about which they will asked to give evidence. They should be properly briefed on the relevant part of the case and they should be cross-examined as early as possible (preferably before pleadings) to ensure that their recollection of facts is consistent with the case pleaded. Considerable harm can be done if pleadings have been exchanged, only to find out a few weeks before the hearing that an important allegation is not supported by facts which come to light during cross-examination of a witness.

Expert witnesses may be called to give evidence on technical matters or on the quantum of a claim. The arbitrator may limit the number of experts to be called. The chosen expert may have played a part in the presentation of the claim, in which case some of the arguments and amounts claimed may have been those put forward by the expert. If this is the case, care should be taken to ensure that the expert addresses his mind to every issue which is open to alternative arguments or methods of calculation. For example, the expert may be fully convinced that the records and facts are sufficient for him to stand firmly by his view of rates for variations or the costs of prolongation. In these circumstances, his evidence on these issues may be valuable at the hearing. On the other hand, if there are concurrent delays, or if he has quantified the cost of disruption, there are bound to be ranges within which the probable cost would fall. In these circumstances, the expert would be abusing the process if he attempted to stand firmly by calculations which were at the extreme end of the range which favoured the party putting him forward as an expert.

If an expert is to command respect and maintain credibility and integrity, he must resist any pressure from his employer, or from his employer's legal advisers, to advance opinions which he does not truly hold. An expert should advance the same opinion whichever party he was representing and this should be tested in 'mock cross-examination' before the hearing. If there is any doubt about the expert's integrity and ability to stand up to cross-examination, he should be withdrawn.

Proofs of evidence by witnesses of fact and expert witnesses may be exchanged before the hearing. This can be useful, particularly if it is used as a means to agree facts and figures before the hearing commences.

The hearing

The hearing often follows similar lines to court proceedings except that they are normally less formal. They are normally held at a neutral venue, such as a hotel, but there is no reason why they should not be held at the offices of one of the parties. The arbitrator formally opens the hearing, followed by:

- The opening address given by the claimant which sets out the issues, the evidence supporting the claimant's case and any submissions on the law which may be relevant;
- Presentation of claimant's witnesses; examination of witnesses on oath by the claimant;
- Cross-examination of claimant's witnesses by the respondent;
- Re-examination of claimant's witnesses by claimant;
- Respondent's opening address;
- Presentation of respondent's witnesses; examination of respondent's witnesses by respondent;
- Cross-examination of respondent's witnesses by claimant;
- Re-examination of respondent's witnesses by respondent;
- Respondent's closing address;
- Claimant's closing address.

The hearing may take one or two days, or it may consist of several hearings over several months. Some hearings may deal with particular issues in dispute, and some may deal with purely procedural matters.

The award

The arbitrator will usually reserve judgement until some weeks after the hearing. The rules governing the arbitration may contain a time limit within which the award must be given. The award is final and binding on the parties, subject to a limited right of appeal pursuant to Section 16 of the Arbitration Act of 1950. The parties may enter into an exclusion agreement under Section 3 of the 1979 Arbitration Act, in which case there can be no review of the award (save for certain exceptions under Section 4 of the Act).

The power to award costs is given by Section 18 of the Arbitration Act of 1950. Where there is only partial success and/or where there are partially successful counter-claims, the apportionment of costs may be complicated. In simple cases, the award of costs is normally in favour of the successful party. However, the conduct of the parties may be taken into account when awarding costs. If an offer of settlement is made during the course of the arbitration, this may be taken into account when awarding costs. In *Tramountana Armadora SA* v. *Atlantic Shipping Co., SA* [1978] 2 All ER 870, the court determined that if the claimant receives no more in the arbitration award than it was offered by the respondent before the award, then costs are assessed against the claimant.

In complex cases, the proceedings may be almost as formal as court proceedings. However, as arbitration is intended to be a relatively quick and inexpensive means of settling disputes, the parties should consider every means of simplifying the manner in which the issues are put before the arbitrator. The following quotations should be taken seriously:

> 'One of the reasons for going to arbitration is to get rid of the technical rules of evidence and so forth.' – Lord Denning in *GKN Centrax Gears Ltd.* v. *Malbro Ltd.* [1965] 2 Lloyds LR 555.

> 'It will be observed that on this occasion the arbitration machinery of the association operated with commendable speed. That may have been because no lawyers were involved.' – *Michael I. Warde* v. *Feedex International, Inc.* [1984] 1 Lloyds LR 310.

Whatever the means of settling disputes, the party who has administered the contract properly, and kept good records, will be much better placed to obtain a favourable result than the party who has barely managed to comply with the basic requirements of the contract.

Appendix: Sample claim for extension of time and additional payment

Introduction to the example

The sample claim which follows is for an extension of time and reimbursement of loss and/or expense arising out of the delays (D1), (D2), (D3) and (D4) shown in Figure 5.8 in Chapter 5. Phased completion has been introduced into the example as a result of which additional activities have become critical.

For simplicity, the claim deals with the subject matter in the main narrative. In practice, particularly for complex claims dealing with many issues, more use would be made of appendices (summarising notices of delay and the like). Copies of relevant correspondence (referred to in the claim), supporting documents, particulars and detailed calculations would also normally be given in an appendix. This example does not contain such appendices (except for programmes and illustrations) but it is assumed that they are submitted.

In this example, clauses referred to in the form of contract are often paraphrased. It is sometimes more appropriate to quote the clauses verbatim.

The claim submission

Covering letter from Better Builders Ltd (the contractor) to T. Square (the architect):

Date 31 March 1992

Dear Sir,

Re: ABC Stores and Depot, New Road, Lower Hamstead, Wilton

Further to our letter of 20 August 1991 requesting a review of extensions of time, our letter of 10 September 1991 giving particulars of loss and/or expense and our letter of 9 February 1992 requesting a copy of the draft final account, to which we have had no response, we enclose herewith our claim for extensions of time, reimbursement of loss and/or expense and damages.

Please note that the contents of this submission do not contain any particulars (with the exception of rates for finance charges for the period after 10 September 1991) which have not been submitted to you previously in correspondence referred to therein. It is our understanding that you have all information necessary for the preparation of the final account and we can see no reason why it should not have been issued prior to this letter.

Our claim is for further extensions of time of two weeks for section A and the works (up to the dates of practical completion) and for reimbursement of loss and/or expense and/or damages for the amount of £60 867.52 (including finance charges on liquidated damages).

We are also requesting the issuance of a certificate of making good defects, a statement pursuant to clause 30.6.1 of the contract (including all adjustments mentioned in the submission), release of retention of £21 010.00, release of liquidated damages amounting to £63 000.00 and a final certificate pursuant to clause 30.8 of the contract.

Your early response would be appreciated.

Yours faithfully
For and on behalf of Better Builders Ltd

Better Builders Ltd
Scaffold Road
Hamstead Rise, Wilton

Manufacturing plant and associated works at
New Road, Lower Hamstead, Wilton
for
ABC Industries Ltd
Factory Lane, Hamstead Rise, Wilton

Claim for extensions of time and
reimbursement of loss and/or expense
and/or damages and repayment of
liquidated damages

Architect: T. Square of Drawing Board and Associates
Design Avenue, Hamstead Rise, Wilton

31 March 1992

Claim for extensions of time for completion of the works and section A, reimbursement of loss and/or expense and/or damages and repayment of liquidated damages.

1.0 **Introduction**.

1.1 **The parties**.

1.1.1 The employer is ABC Industries Ltd of Factory Lane, Hamstead Rise, Wilton.

1.1.2 The architect is T. Square of Drawing Board and Associates, Design Avenue, Hamstead Rise, Wilton.

1.1.3 The quantity surveyor is R. E. Measure of The Manor, Billingsgate Road, Hamstead Rise, Wilton.

1.1.4 The contractor is Better Builders Ltd of Scaffold Road, Hamstead Rise, Wilton.

1.2 **The works**.

1.2.1 The works comprise the alteration of an existing stores building into a manufacturing plant for motor parts including the construction of a new access road, drainage, diversion of services and landscaping at ABC Stores and Depot, New Road, Lower Hamstead, Wilton.

1.3 **The tender and the contract sum**.

1.3.1 The contractor submitted his tender on 10 January 1991 for the sum of £827 333.00. It was a condition of the contractor's tender that work would be permitted on weekends and public holidays and that the employer would undertake to ensure the presence of the architect or his representative on such days where it was necessary for the supervision and administration of the contract.

1.3.2 The employer unconditionally accepted the contractor's tender by letter dated 22 January 1991.

1.3.3 The contract sum in article 2 of the agreement is £827 333.00.

1.4 **The contract.**

1.4.1 The contract is the Standard Form of Building Contract, 1980 Edition, incorporating amendments 1,2,4,5 and 6, Private Edition with Quantities, issued by the Joint Contracts Tribunal and incorporating the Sectional Completion Supplement revised January 1989. The following amendments have been made to the standard conditions of contract:

1.4.1.1 Sub-clause 1.3 – Definitions.
Definition of Section A – 'Completion of all alterations in the existing store building to such state as (in the opinion of the architect) to enable the employer to commence installation of plant and equipment.'

1.4.1.2 Sub-clause 25.4.2 (relevant event – exceptionally adverse weather conditions) has been deleted.

1.4.2 The relevant particulars in the appendix to the contract are as follows:

1.4.2.1 Clause 1.3 Dates for completion
– Twenty-two weeks after the date of possession.

1.4.2.2 Clause 17.2 Defects liability period
– Six months.

1.4.2.3 Clause 22.1 Insurance of the works
– Alternative C applies.

1.4.2.4 Clause 23.1.1 Date of possession
– Seven days after the architect's written instruction to take possession of the site.

1.4.2.5 Clause 23.1.2 Deferment of the date of possession
– Does not apply.

1.4.2.6 Clause 22.2 Liquidated and ascertained damages
– £2500.00 per day.

1.4.2.7 Clause 30.4.1.1 Retention percentage
– Five per cent.

1.4.2.8 Clauses 38, 39 and 40 Fluctuations
– Clause 38 shall apply.

1.4.3 The relevant particulars in the appendix to the sectional completion supplement are as follows:

1.4.3.1 Clause 2.1 Section of the works

– Section A as described in clause 1.3 of the conditions of contract.

1.4.3.2 Clause 18.1.5 Section value
– £525 000.00.

1.4.3.3 Clauses 17, 18, 30 Defects liability period
– Six months.

1.4.3.4 Date of possession of section
– On the date of possession in clause 23.1.1 of the conditions of contract.

1.4.3.5 Date for completion of section
– Sixteen weeks after the date of possession.

1.4.3.6 Rate of liquidated and ascertained damages for section
– £2000.00 per day.

1.5 **The programme:**

1.5.1 The contractor's original programme for completion of the works is shown in appendix I hereto (see Figure A1).

1.5.2 The activities forming section A are F-G, B-G and G-H.

2.0 **Summary of Facts:**

2.1 **Possession of site: commencement and completion of the works.**

2.1.1 On 4 February 1991, the architect gave written notice to the contractor to take possession of the site on 11 February 1991.

2.1.2 The contractor took possession of the site and commenced work on 11 February 1991.

2.1.3 Pursuant to clause 3.1 of the conditions of contract, the sectional completion supplement (and the relevant appendices) and the architect's written instruction of 4 February 1991, the dates for completion were:

2.1.3.1 Section A – 2 June 1991.

2.1.3.2 The works – 14 July 1991.

2.1.4 Practical completion occurred on the following dates:

2.1.4.1 Section A – 3 June 1991. (Architect's certificate of practical completion dated 9 August 1991).

2.1.4.2 The works – 4 August 1991. (Architect's certificate of practical completion dated 9 August 1991).

2.2 **Delay and extensions of time:**

2.2.1 The contractor gave the following notices of delay and particulars pursuant to clause 25 of the conditions of contract:

2.2.1.1 Letter dated 18 March 1992 [week6] – Notice of delay as a result of exceptionally adverse weather conditions affecting activity B-E (delay D1).

2.2.1.2 Letter dated 21 March 1991 [week 6] – Notice of delay as a result of architect's instruction no 1 (issued 18 March 1991) to alter work partially completed to activity B-G (delay D2).

2.2.1.3 Letter dated 9 April 1991 [week 9] – Particulars of delay caused by architect's instruction no 1.

2.2.1.4 Letter dated 2 April 1991 [week 8] – Notice of delay as a result of revised and additional work to activity B-G shown on drawings AD/14A and AD/15A issued on 1 April 1991 [week 8] (delay D3).

2.2.1.5 Letter dated 26 June 1991 [week 20] – Particulars of delay caused by the issuance of drawings AD/14A and AD/15A.

2.2.1.6 Letter dated 10 July 1991 [week 22] – Notice of delay as a result of late issuance of instructions on the expenditure of the P C sum for work to be done by a nominated subcontractor on activity H-K (delay D4).

2.2.1.7 Letter dated 5 August 1991 – Particulars of delay caused by late issuance of instructions on the expenditure of PC sum (see 2.2.1.6 hereof).

2.2.1.8 Letter dated 20 August 1991 – Letter requesting the architect to review his extensions of time for section A and the works pursuant to clause 25.3.3 of the conditions of contract and giving further particulars.

2.2.2 The architect has made the following extension of time for completion of the works pursuant to clause 25 of the conditions of contract:

2.2.2.1 Certificate reference EOT 1 dated 12 August 1991 [week 27] Section A – Extension of time of one week as a result of the additional work to activity B-G shown on drawings AD/14A and AD/15A (delay D3), giving a revised completion date of 9 June 1991.

2.2.2.2 Certificate reference EOT 2 dated 12 August 1991 [week 27] The works – Extension of time of one week as a result of the late issuance of instructions for the expenditure of PC sums (delay D4), giving a revised completion date of 21 July 1991.

2.2.2.3 At the date of this submission, the architect has not given a written response to the contractor's request of 20 August (see 2.2.1.8 hereof).

2.3 **Certificates of non-completion.**

2.3.1 Pursuant to clause 24.1 of the conditions of contract, the architect issued certificates of non-completion dated 12 August 1991 certifying that the contractor had not completed:

Section A – by the extended date of completion of 9 June 1991.

The works – by the extended date of completion of 21 July 1991.

2.4 **Direct loss and/or expense:**

2.4.1 The contractor notified the architect, pursuant to clause 26 of the conditions of contract, that the regular progress of the works had been affected and that he had incurred, and was continuing to incur, direct loss and/or expense as follows:

2.4.1.1 Letter dated 28 May 1991 [week 16] – As a result of delays to activity B-G (delays D2 and D3).

2.4.1.2 Letter dated 25 June 1991 [week 20] – Further disruption of the regular progress of the works as a result of delay to activity B-G (delay D3) and as a result of late nomination of the subcontractor for activity H-K (delay D4).

2.4.2 By letter dated 12 August 1991, the quantity surveyor requested further particulars from the contractor in support of his application for reimbursement of direct loss and/or expense.

2.4.3 On 10 September 1991, the contractor provided the further particulars requested by the quantity surveyor on 12 August 1991.

2.4.4 At the date of this submission, no sums for loss and/or expense have been ascertained and no further requests for particulars have been made by the architect or quantity surveyor.

2.5 **Payment and final account:**

2.5.1 The latest certificate issued prior to the date of this submission is interim payment certificate no 6 dated 12 August 1991 showing the following amounts:

2.5.1.1	Gross value of work at practical completion	£840 400.00.
2.5.1.2	Retention	£21 010.00.
2.5.1.3	Nett amount due	£819 390.00.
2.5.1.4	Previous certificates	£725 200.00.
2.5.1.5	Amount due for payment	£94 190.00.

2.5.2 The employer has paid the amount certified as being due for payment in interim payment certificates, less liquidated damages in the sum of £63 000.00. The nett payment made after deduction of liquidated damages was £31 190.00.

2.5.3 On 9 February 1992, the contractor requested a copy of the final account showing the value of work executed including all adjustments to the contract sum and amounts for nominated subcontractors and suppliers.

2.5.4 At the date of this submission, no final account has been issued to the contractor.

2.6 **Defects:**

2.6.1 On 5 January 1992, the architect issued a schedule of defects pursuant to clause 17.3 of the conditions of contract and instructed the contractor to make good the said defects.

2.6.2 On 9 February 1992,the contractor notified the architect that he had rectified all defects notified by the architect in his schedule of 5 January 1992 and he requested a certificate of making good defects pursuant to clause 17.4 of the conditions of contract.

2.6.3 At the date of this submission, no certificate of making good defects has been issued.

3.0 **Basis of claim:**

3.1 The contract contained the following provisions:

3.1.1 Clause 13.5 – If compliance with an instruction substantially changes the conditions under which any other work is executed, then such work shall be treated as if it had been the subject of an instruction of the architect requiring a variation under clause 13.2. Provided that no allowance shall be made under clause 13.5 for any effect on the regular progress of the works or for any other direct loss and/or expense for which the contractor would be reimbursed by payment under any other provisions in the conditions of contract.

3.1.2 Clause 17.4 – When in the opinion of the architect any defects or other faults which he may have required to be made good under clauses 17.2 and 17.3 (defects occurring in the defects liability period), he shall issue a certificate to that effect and the said defects shall be deemed to have been made good on the day named in such certificate.

3.1.3 Clause 24.2.2 – If, under clause 25.3.3, the architect fixes a later completion date the employer shall repay to the contractor liquidated damages allowed under clause 24.2.1 for the period up to such later completion date.

3.1.4 Clause 25 – The contractor shall give notice and particulars of delay and shall be entitled to a fair and reasonable extension of time for completion if completion of the works (and/or section) are delayed by the following relevant events (specified in clause 25.4);

3.1.4.1 – compliance with architect's instructions under clauses 13.2 (variations) – clause 25.4.5.1;

3.1.4.2 – the contractor not having received in due time necessary instructions for which he specifically applied in writing provided that such application was made on a date having regard to the completion date was neither unreasonably distant from nor unreasonably close to the date when it was necessary to receive the same (clause 25.4.6).

3.1.5 Clause 26 – If the contractor makes written application to the architect stating that he has incurred or is likely to incur direct loss and/or expense for which he would not be reimbursed under any other provision in the contract due to the regular progress of the works or any part thereof being materially affected by:

3.1.5.1 – the contractor not having received in due time necessary instructions for which he specifically applied in writing provided that such application was made on a date having regard to the completion date was neither unreasonably distant from nor unreasonably close to the date when it was necessary to receive the same (clause 26.2.1);

3.1.5.2 – architect's instructions issued under clause 13.2 requiring a variation (clause 26.2.7);

and providing that his application was made as soon as possible after it has become, or should reasonably have become, apparent to the contractor that the regular progress of the works or any part thereof had been or is likely to be affected,

and the contractor has in support of his application upon the request of the architect submitted such information as should reasonably be necessary to enable the architect to form an opinion, and

the contractor has submitted to the architect or quantity surveyor upon request such details of loss and/or expense as are reasonably necessary for ascertainment,

then the architect or the quantity surveyor shall ascertain the amount of such loss and/or expense and the amount ascertained shall be added to the contract sum (clauses 26.1 and 26.5).

3.1.6 Clause 30 – Half of the retention percentage may be deducted from the amount which relates to work which has reached practical completion (clause 30.4.1.3) and the remaining half shall be released upon issuance of the final certificate, which shall be issued no later than two months after whichever of the following occurs last (clause 30.8):

3.1.6.1 the end of the defects liability period;

3.1.6.2 the date of the issue of the certificate of making good defects under clause 17.4;

3.1.6.3 the date on which the architect sent a copy to the contractor of any ascertainment under clause 30.6.1.2.1 (loss and/or expense) or statement under clause 30.6.1.2.2 (all adjustments to the contract sum).

3.2 The above provisions apply to the works and sections A (sectional completion supplement).

3.3 *Without prejudice* to the contractor's rights to claim damages under the general law (clause 26.6), save as provided in 3.3.1 and 3.3.2 hereof, the contractor's claim is made pursuant to the provisions on the contract hereinbefore mentioned.

3.3.1 The contractor is entitled to interest on liquidated damages which shall become repayable to the contractor pursuant to a revised extension of time made by the architect – *Department of Environment for Northern Ireland* v. *Farrans* (1981) 19 BLR 1.

3.3.2 Where the contractor complies with his obligations with respect to information and particulars for the purposes of preparing the final account and all adjustments to be made to the contract sum, if the architect or quantity surveyor fail to prepare such final account or make all necessary adjustments as aforesaid, the contractor is entitled to reimbursement of the cost incurred in preparing such adjustments – *James Longley & Co Ltd* v. *South West Regional Health Authority* (1985) 25 BLR 56.

4.0 **Details of Claim:**

4.1 **Introduction.**

4.1.1 The contractor's programme for completion of the works and section A within the periods for completion is shown in appendix I (A1) hereto. Activities A-B to J-K are critical for completion of the works in twenty-two weeks. Activities A-B to E-F, F-G and G-H are critical for completion of section A in sixteen weeks. Activities B-C to D-H and H-K are not critical, and will not become critical until all of the float shown on the contractor's programme has been used up by delays to these otherwise non-critical activities.

4.1.2 The causes of delay referred to in this section are delays which entitle the contractor to an extension of time, or, if no extension of time is permitted for delay by such cause (as in the case of exceptionally adverse weather conditions), the contractor would be entitled to an extension of time for other causes of delay which used the float in the programme as a result of which otherwise non-critical activities became critical and caused delay to completion of the works (or section).

4.2 **Exceptionally adverse weather conditions – delay (D1).**

4.2.1 Activity B-E is for the construction of a surface water culvert under the new access road.

4.2.2 The contractor completed the preceding activity (A-B) on programme and was proceeding with the construction of activity B-E in accordance with the programme.

4.2.3 During the week-end of 16 and 17 March 1991, continuous rainfall caused the open trench for the construction of the culvert to be flooded. On 18 March 1991, the contractor hired additional pumps to remove the water from the excavations. However, exceptionally adverse weather conditions continued during the period of two weeks (weeks commencing 18 and 25 March 1991). Records of the rainfall during the period taken at Much Hamstead (four miles from the site) were obtained by the architect for record purposes.

4.2.4 Water had been removed from the trenches and the contractor was able to recommence construction of the culvert on 1 April 1991 (a delay of two weeks).

4.2.5 The contractor gave notice of delay pursuant to clause 25.2.2.1 of the conditions of contract.

4.2.6 It is common ground that the contractor was delayed by a period of two weeks as a result of the said weather conditions and that no extension of time is permitted for such delay by virtue of the deletion of clause 25.4.2 of the conditions of contract.

4.3 **Architect's instruction no 1 – delay (D2).**

4.3.1 Activity F-G is for the construction of an effluent drain under the existing stores and constructing new bases for the plant and equipment to be installed by the employer.

4.3.2 On 18 March 1991, the architect issued instruction no 1 which required the contractor to reposition the effluent drain in order to accommodate foundations for future alterations to the stores by the employer.

4.3.3 At the time of issuance of the said instruction, the construction of the new effluent drain was on programme. The contractor had excavated and laid all pipes within the existing stores and was ready to test the pipes prior to backfilling the trench on 18 March 1991. Records of the work executed prior to the issuance of the said instruction were agreed with the quantity surveyor.

4.3.4 The contractor commenced cutting out the existing floor slab at the revised location of the effluent drain on 19 March 1991. On the same day, some of the resources (labour and plant) were diverted from activity B-E (delayed as a result of the inclement weather described in 4.2 hereof) to commence backfilling to the redundant length of effluent drain.

4.3.5 The contractor excavated the trench for the revised effluent drain and laid the pipes and was ready for testing on 1 April 1991. A delay of two weeks had occurred as a result of the said instruction. The time taken to carry out the work prior to testing (2 weeks) was the same time allowed in the contractor's programme for carrying out the same quantity of work in the originally designed location of the effluent drain.

4.3.6 Backfilling and making good the floor slab at the location of the redundant effluent drain was completed on 1 April 1991. Had the contractor not been able to utilise resources from activity B-E (see 4.3.4 hereof), this work could not have been executed until after the contractor had completed the diversion of the effluent drain to the revised location.

4.3.7 As a result of the foregoing, activity B-G had been delayed by two weeks. No direct delay to completion of section A or the works was caused by the said instruction – see appendix II (A2) hereto.

4.3.8 Notices and particulars of the delay and disruption and loss and/or expense caused by the said instruction were given by the contractor pursuant to clauses 25 and 26 of the conditions of contract (see 2.2 and 2.4 hereof).

4.4 **Additional work – Delay (D3):**

4.4.1 On 1 April 1991, the contractor notified the architect, in writing (letter ref BB/10), that he intended to divert resources from activity B-G in order to make up the time lost due to exceptionally adverse weather conditions (delay D1). The contractor's revised programme showing completion by the original completion date was attached to the said letter – see appendix II – (A3) hereto. The revised programme was made on the basis of using some of the float on activity B-G. The original float of six weeks had been reduced by two weeks (delay D2) and the contractor envisaged using two weeks of the remaining four weeks float so that work could cease on activity B-G until such time as activity B-E was on programme. No delay to completion of section A or the works would occur as a result of the re-programming and two weeks; float would remain in activity B-G.

4.4.2 On 1 April 1991, the architect issued drawings AD/14A and 15A showing four additional bases for machinery (to be installed by the employer) and additional effluent branch drains.

4.4.3 On 2 April 1991, the contractor had set out for the new bases and ordered materials for the additional work. On the same day the contractor notified the architect that he estimated a delay of seven to eight weeks to activity B-G as a result of the said instruction (see 2.2.1.4 hereof). In the same letter, the contractor notified the architect that it would not be of any benefit to divert resources from activity B-G to activity B-E (see 4.4.1 hereof) as completion of section A was dependent upon the timely completion of activity B-G, which had now become critical as a result of the additional work.

4.4.4 The contractor had completed all work to the revised drawings, by 16 June 1991 (a delay of 7 weeks).

4.4.5 On 18 June 1991 [week 19], the contractor issued his revised programme showing the delays D1 to D3, completion of section A on 23 June 1991 [end of week 19] and completion of the works on 28 July 1991 [end of week 24] – see appendix II (A4) hereof.

4.4.6 Notices and particulars of the delay and disruption and loss and/or expense caused by the said additional work were given by the contractor pursuant to clauses 25 and 26 of the conditions of contract (see 2.2 and 2.4 hereof).

4.5 **Late instruction for expenditure of PC sum – Delay (D4).**

4.5.1 The contract bills included the PC sum £45 000.00 for the supply and installation of mechanical equipment to the effluent treatment plant. This was shown on the contractor's original programme as activity H-K commencing in week 19 and the period for installation was one week.

4.5.2 The contractor's covering letter submitted with the said programme indicated that approximately four weeks would be necessary for ordering, manufacture and delivery of standard equipment from several well-known firms. The letter went on to request the architect to notify the contractor in the event of any potential subcontractors requiring a longer period for delivery, manufacture or installation. The necessary instructions (for standard equipment) would be required no later than 20 May 1991 (commencement of week 15).

4.5.3 As a result of delays D2 and D3 (see 4.3 and 4.4 hereof) the revised latest date for receipt of instructions was 3 June 1991 [week 17].

4.5.4 On 3 June 1991, the architect issued instruction no 7 for the supply and installation of the equipment to be done by Pumps & Co for the sum of £42 250.00 in accordance with the tender documents attached to the said instruction. The delivery period for the equipment (which was not a standard set) was quoted as seven to eight weeks and one week was required for installation.

4.5.5 On the same day, the contractor notified the architect by 'fax (ref BB/77) that the delivery period quoted by Pumps & Co was unacceptable, but he would be prepared to place the order with Pumps & Co provided that the architect would make an appropriate extension of time.

4.5.6 On 4 June 1991, the architect notified the contractor by 'fax (ref TS/12A) that he would take the delivery period of the pumps into account when making his decision on extensions of time.

4.5.7 On 5 June 1991, the contractor placed his order with Pumps & Co. A formal subcontract was signed between the contractor and Pumps & Co on 17 June 1991.

4.5.8 Pumps & Co delivered their equipment to site on 29 July 1991 and completed the installation, including testing, on 4 August 1991 [end of week 25]. Completion of the works had been delayed by three weeks having regard to the fact that the contractor had been denied the opportunity to reduce the delay cause by exceptionally adverse weather conditions (delay D1 – see 4.2 and 4.4.1 hereof) – see appendix II (A5) hereto.

4.5.9 Notices and particulars of the delay and disruption and loss and/or expense caused by the said additional work were given by the contractor pursuant to clauses 25 and 26 of the conditions of contract (see 2.2 and 2.4 hereof).

4.6 **Summary:**

4.6.1 Completion of section A has been delayed by three weeks as a result of delays (D2) and (D3) – (see 4.3 and 4.4).

4.6.2 Completion of the works has been delayed by three weeks as a result of delays (D2), (D3) and (D4) – (see 4.3, 4.4 and 4.5 hereof).

4.6.3 The delays referred to hereinbefore are shown in appendix II (A5) hereto.

4.6.4 The contractor contends that the architect has wrongly deducted the period of two weeks (delay caused by exceptionally adverse weather conditions) from the total delay to completion of three weeks for section A and the works. (The architect's reasons for making this adjustment are given in minutes of meeting of 12 August 1991, paragraph 2.3).

4.6.5 Even if the contractor had not contemplated reprogramming the works to mitigate the delay (D1) – (see paragraph 4.4.1 hereof), the contractor maintains that no deduction should be made for delay (D1) when, in any event, completion of section A and the works were delayed by delays (D2), (D3) and (D4) which were the responsibility of the employer. Accordingly, the employer could not levy liquidated damages for the period of two weeks when the progress of the works was delayed by matters for which the employer was responsible.

4.6.6 Further, or alternatively, the contractor was prevented from mitigating the delay (D1) as a result of the additional work (see 4.4 hereof) and is entitled to a fair and reasonable extension of time of three weeks pursuant to clause 25 of the conditions of contract (relevant events described in clauses 25.4.5.1 and 25.4.6) until the date of practical completion of section A and the works and for reimbursement of loss and/or expense pursuant to clause 26 of the conditions of contract (matters described in clause 26.2.1 and 26.2.7).

5.0 **Evaluation of Loss and/or Expense:**

5.1 For the reasons given in 4.0 hereof, the contractor is entitled to direct loss/and or expense as follows:

5.1.1 **Prolongation:**
The period of prolongation is 3 weeks. The contractor contends that the issuance of drawings AD/14A and 15A (see 4.3 hereof) substantially changed the conditions under which the work on activity B-E would otherwise have been carried out (see 4.4.1 hereof). Therefore, notwithstanding delay (D1), pursuant to the provisions of clause 13.5.5 and the proviso in the final paragraph of clause 13.5, the contractor is entitled to reimbursement for the total period of prolongation pursuant to clause 26 (matter referred to in clause 26.2.7).

The contractor is entitled to reimbursement of loss and/or expense caused by delays (D2) and (D3) pursuant to clause 26 (matter described in clause 26.2.7).

The contractor is entitled to reimbursement of loss and/or expense caused by delay (D4) pursuant to clause 26 (matter described in clause 26.2.1).

5.1.1.1 **Head office overheads and profit:**
As a result of the delays (D2), (D3) and (D4) described in 4.0 hereof, the contractor was required to retain its key staff and resources on site for an additional period of three weeks and was deprived of making a contribution to overheads and profit. The contractor is therefore entitled to recover this loss pursuant to the provisions mentioned in 5.1.1 hereof.

The contractor's auditors have certified that the contractor's overheads and profit (as percentages of revenue) were as follows:

Year ending 31 July 1990 – 12.76 %
Year ending 31 July 1991 – 11.98 %

The average percentage for overheads and profit for two years was therefore:

(12.76 + 11.98)/2 = 12.37 %

Using *Emden's* formula:

Loss of overheads and profit for three weeks =

$$\frac{\text{Overheads \& profit \%}}{100} \times \frac{\text{contract sum}}{\text{contract period}} \times \text{period of delay}$$

$$\frac{12.37\%}{100} \times \frac{\text{£827 333.00}}{\text{22 weeks}} \times \text{3 weeks} = \quad \textbf{£13 955.00}$$

5.1.1.2 **Site overheads and establishment (preliminaries):**
As a result of the delays (D1), (D2) and (D3) described in 4.0 hereof, the contractor was required to retain its key staff and resources on site for an additional period of three weeks. The contractor is therefore entitled to recover the expense of his site overheads and establishment costs for the period of delay pursuant to the provisions mentioned in 5.1.1 hereof.

Delays (D2) and (D3) – 2 weeks – see (A4) in appendix II hereto.

Costs incurred during weeks 11 and 12;

Excludes costs associated with activity B–G:

Project manager	2 weeks @ £675.00/week	= £1350.00
General foreman	2 weeks @ £565.00/week	= £1130.00
Engineer	2 weeks @ £550.00/week	= £1100.00
Quantity surveyor (part)	2 weeks @ £310.00/week	= £620.00
Administration staff	2 weeks @ £388.00/week	= £776.00
Hire of offices	2 weeks @ £455.00/week	= £910.00
Office equipment	2 weeks @ £105.00/week	= £210.00
Plant & equipment	2 weeks @ £967.00/week	= £1934.00
Scaffolding	2 weeks @ £761.00/week	= £1522.00
Small tools & equipment	2 weeks @ £325.00/week	= £650.00
Electricity charges	£1430.00 × 2/13 weeks	= £220.00
Telephone charges	£650.00 × 2/13 weeks	= £100.00
Security	2 weeks @ £250.00/week	= £500.00
Stationery and sundries	£90.00 × 14/30 days	= £42.00
Total		£11 064.00

Delay (D4) – One week – see (A5) in appendix II hereto.

Costs incurred during week 23;

Project manager	1 week @ £675.00/week	= £675.00
General foreman	1 week @ £565.00/week	= £565.00
Quantity surveyor (part)	1 week @ £310.00/week	= £310.00
Administration staff	1 week @ £153.00/week	= £153.00
Hire of offices	1 week @ £455.00/week	= £455.00
Office equipment	1 week @ £105.00/week	= £105.00
Plant & equipment	1 week @ £275.00/week	= £275.00
Small tools & equipment	1 week @ £120.00/week	= £120.00
Electricity charges	£650.00 × 1/13 weeks	= £50.00
Telephone charges	£325.00 × 1/13 weeks	= £25.00
Security	1 week @ £250.00/week	= £250.00
Stationery and sundries	£62.00 × 7/31 days	= £14.00
Total		£2997.00

Total site overheads and establishment costs = £11 064.00 + £2997.00
= **£14 061.00**

5.1.1.3 **Finance charges on delayed release of retention:**
Pursuant to clauses 30.4 and 30.8 of the conditions of contract, two-and a half per cent of the contract sum (being one half of the retention percentage stated in the appendix to the conditions of contract) should be certified and paid after practical completion (of section A and the works) and upon the issuance of the final certificate.

As a result of the delays (D2), (D3) and (D4), the dates when the retention ought to have been released were three weeks later than the dates which would have applied if there had been no delay. Accordingly, the contractor has incurred financing charges by virtue of the fact that interest charges on his overdraft have been accruing for an additional period of three weeks on the amount of retention withheld.

The finance charges incurred are calculated at the rate of two per cent above the bank base rate (as charged by the contractor's bank from time to time) as follows:

First half due to be released.

Period of financing (assume release three weeks after practical completion):

	Planned release	actual release	Rate
Section A	8 July 1991	29 July 1991	13 %
The works	5 August 1991	26 August 1991	13 %

Amount of retention:

Section A –£14 000.00

Finance charges = £14 000.00 × 13 % × 21/365= £104.71

The works –£21 010.00 – £14 000.00 = £7 010.00

Finance charges = £7 010.00 × 13 % × 21/365= £52.43

Second half due to be released (Defects liability period – six months).

Period of financing (assume release six months after first release):

	Planned release	actual release	Rate
Section A	8 Jan 1992	29 Jan 1992	12.5 %
The works	5 February 1992	26 February 1992	12.5 %

Amount of retention:

Section A –£14 000.00

Finance charges = £14 000.00 × 12.5 % × 21/366= £100.41

The works –£21 010.00 – £14 000.00 = £7 010.00

Finance charges = £7 010.00 × 12.5 % × 21/366= £50.28

Total finance charges on retention

= £104.71 + £52.43 + £100.41 + £50.28= **£307.83**

5.1.1.4 **Fluctuations:**
The contract does not provide for reimbursement of fluctuations of labour or materials (see 1.4.2.8 hereof). The contractor allowed for the anticipated increase in labour in June 1991 in his tender (for the labour required to execute the work in weeks 20–22 on activity J-K). The hours allowed by the contractor in his tender during this period were as follows:

Craft operatives	–3170 hours
Labourers	–2700 hours

Due to delays (D2), (D3) and (D4), the contractor's labour resources in weeks 20–25 were as follows:

Craft operatives	–5060 hours
Labourers	–4365 hours

Due to the fact that the contractor had been prevented from mitigating the delay caused by exceptionally adverse weather conditions (delay D1) – see 4.4.1 hereof, the additional costs of labour for the additional hours expended after the wage increase on 24 June 1991 (most of which would have been prevented by the measures proposed by the contractor to mitigate the delay) qualify for reimbursement pursuant to clause 26 of the conditions of contract.

The additional costs of labour claimed are calculated as follows:

	Tender	24 June 1991	Increase
Craft operatives	£3.38	£3.57	
NI & Employer's Ins. (11%)	£0.37	£0.39	
	£3.75	£3.96	£0.21 (hr)

Labourers	£2.88	£3.03	
NI & Employer's Ins. (11%)	£0.32	£0.33	
	£3.20	£3.36	£0.16 (hr)

Hours after 24 June 1991:

Craft operatives	5060 –	3170	= 1890 hrs
Labourers	4365 –	2700	= 1665 hrs

Therefore, the additional costs caused by delays (D1), (D2) and (D3) are;

Craft operatives	1890 hrs @ £0.21	= £396.90	
Labourers	1665 hrs @ £0.16	= £266.40	
Total			£663.30

The total increased cost of labour fluctuations is **£663.30**

The contractor ordered all materials at the prices applicable at the date of tender and no claim is made for increased costs of materials.

5.1.1.5 **Total prolongation costs:**

Head office overheads & profit (5.1.1.1)	= £13 955.00
Site overheads & establishment costs (5.1.1.2)	= £14 061.00
Finance charges on retention (5.1.1.3)	= £307.83
Fluctuations (5.1.1.4)	=£663.30
TOTAL	**£28 987.13**

5.1.2 **Disruption:**
Activity B-G was delayed by nine weeks as a result of delays (D2) and (D3). Site staff and resources allocated to this activity were required on site for this additional period and the contractor is entitled to reimbursement of expense caused thereby.

5.1.2.1 **Cost of resources allocated to activity B-G:**

Section foreman	9 weeks @ £503.00/week	= £4527.00
Engineer	9 weeks @ £510.00/week	= £4590.00
Plant & equipment	9 weeks @ £300.00/week	= £2700.00
Scaffolding (part only)	5 weeks @ £470.00/week	= £2350.00
Small tools & equipment	9 weeks @ £225.00/week	= £2025.00
Total		**£16 192.00**

5.1.3 **Finance charges on loss and expense:**
The contractor has incurred financing charges by virtue of the fact that interest charges on his overdraft have been accruing from the date that each head of loss and expense occurred.

In addition, the contractor has incurred finance charges on the liquidated damages and he claims finance charges under the general law until liquidated damages are repaid to the contractor (see 3.3.1 hereof).

For the purposes of calculating finance charges, the dates when the loss and expense occurred are taken as follows:

Head office overheads & profit (5.1.1.1)	– £13 955.00	– 1 August 1991
Site overheads & establishment (5.1.1.2)	– £11 064.00	– 1 May 1991
	– £2 997.00	– 1 August 1991
Finance charges on retention (5.1.1.3)	– £104.71	– 1 August 1991
	– £52.43	– 1 Sept 1991
	– £100.41	– 1 Feb 1992
	– £50.28	– 1 March 1992
Disruption (5.1.2.1)	– £16 192.00	– 1 May 1991
Fluctuations (5.1.1.4)	– £663.30	– 1 August 1991
Total	– £45 179.13	
On liquidated damages	– £63 000.00	– 1 Sept 1991

Therefore, finance charges accrued on the following sums from the dates given below:

£27 256.00	– 1 May 1991
£17 720.01	– 1 August 1991
£63 052.43	– 1 September 1991
£100.41	– 1 February 1992
£50.28	– 1 March 1992

The finance charges incurred are calculated at the rate of two per cent above the bank base rate (as charged by the contractor's bank from time to time) in appendix III hereto.

The total finance charges up to 31 March 1992 (the date of this submission) are **£9 638.39**.

5.1.4 **Costs of preparing the claim:**

5.1.4.1 The contractor has complied in all respects with his obligations to give notice and full particulars pursuant to clause 26 of the conditions of contract (see 2.2 and 2.4 hereof) and the architect has failed to comply with his obligations to ascertain the loss and/or expense due to the contractor.

5.1.4.2 Accordingly the contractor claims reimbursement of the fees paid to Contraconsult Ltd for the preparation of this submission in the sum of **£6 050.00** (see 3.3.2 hereof).

5.2 **Summary of loss and/or expense and/or damages;**
The following sums are due to the contractor:

Prolongation costs (5.1.1.5)	– £28 323.83
Disruption (5.1.2.1)	– £16 192.00
Finance charges (5.1.3)	– £9 638.00
Cost of preparing the claim (5.1.4)	– £6 050.00
Total	**–£60 867.52**

6.0 **Statement of Claim;**

6.1 **Extensions of time:**

6.1.1 The contractor claims an extension of time pursuant to clause 25 of the conditions of contract of a further two weeks giving the following extended dates for completion:

Section A– 23 June 1991
The works– 4 August 1991

6.2 **Loss and expense and/or damages:**

6.2.1 The contractor claims reimbursement of loss and/or expense pursuant to clause 26 of the conditions of contract and/or damages for breach of contract amounting to **£60 867.52**.

6.3 **Retention:**

6.3.1 The contractor is entitled to release of retention in the sum of £21 010.00.

6.4 **Adjustments to the contract sum:**

6.4.1 The contractor has submitted under separate cover (letter of even date) his statement of account for all adjustments to the contract sum (excluding the loss and/or expense and/or damages herein) and claims payment of the sum **£6 325.78** being the outstanding amount due to be included in the final statement of account pursuant to clause 30.6.1 of the conditions of contract.

6.5 **Liquidated damages:**

6.5.1 The contractor claims repayment of liquidated damages in full for the amount of £63 000.00.

6.6 **Finance charges accruing:**

6.6.1 The contractor claims finance charges on the sums stated in 6.2 to 6.5 hereof after the date of this submission at the rate of two per cent above the bank base rate.

Appendix I

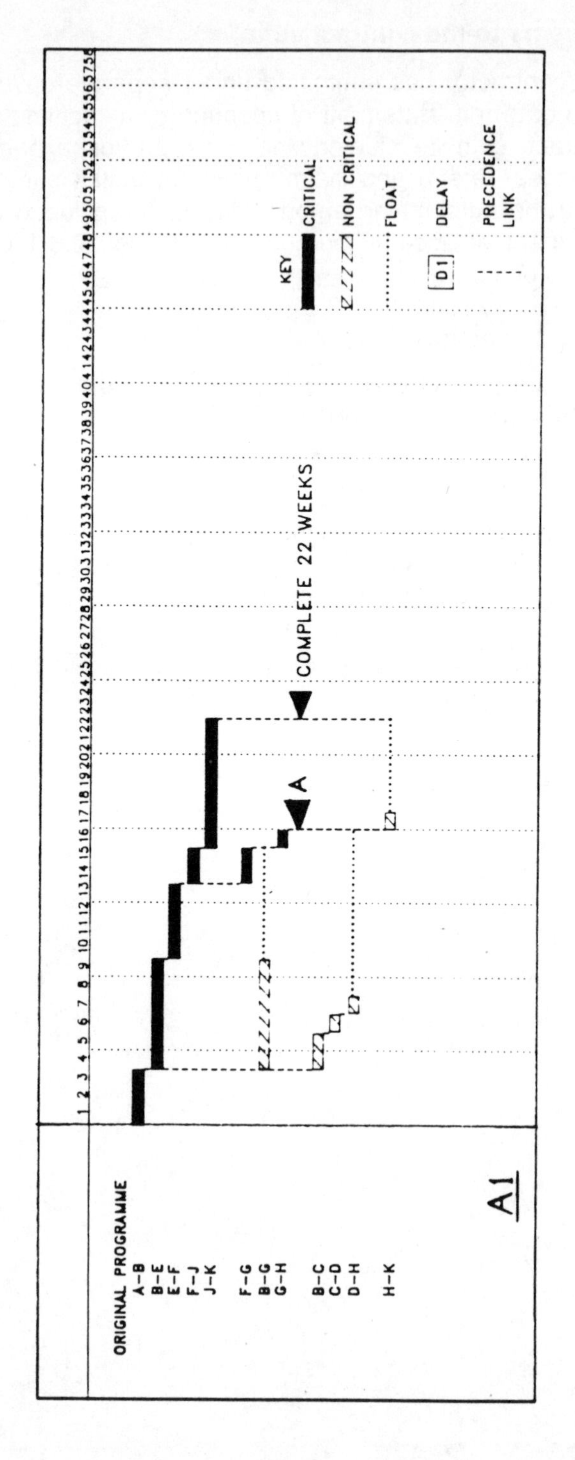

Figure A. 1 Precedence (linked) bar chart – original programme

Appendix II

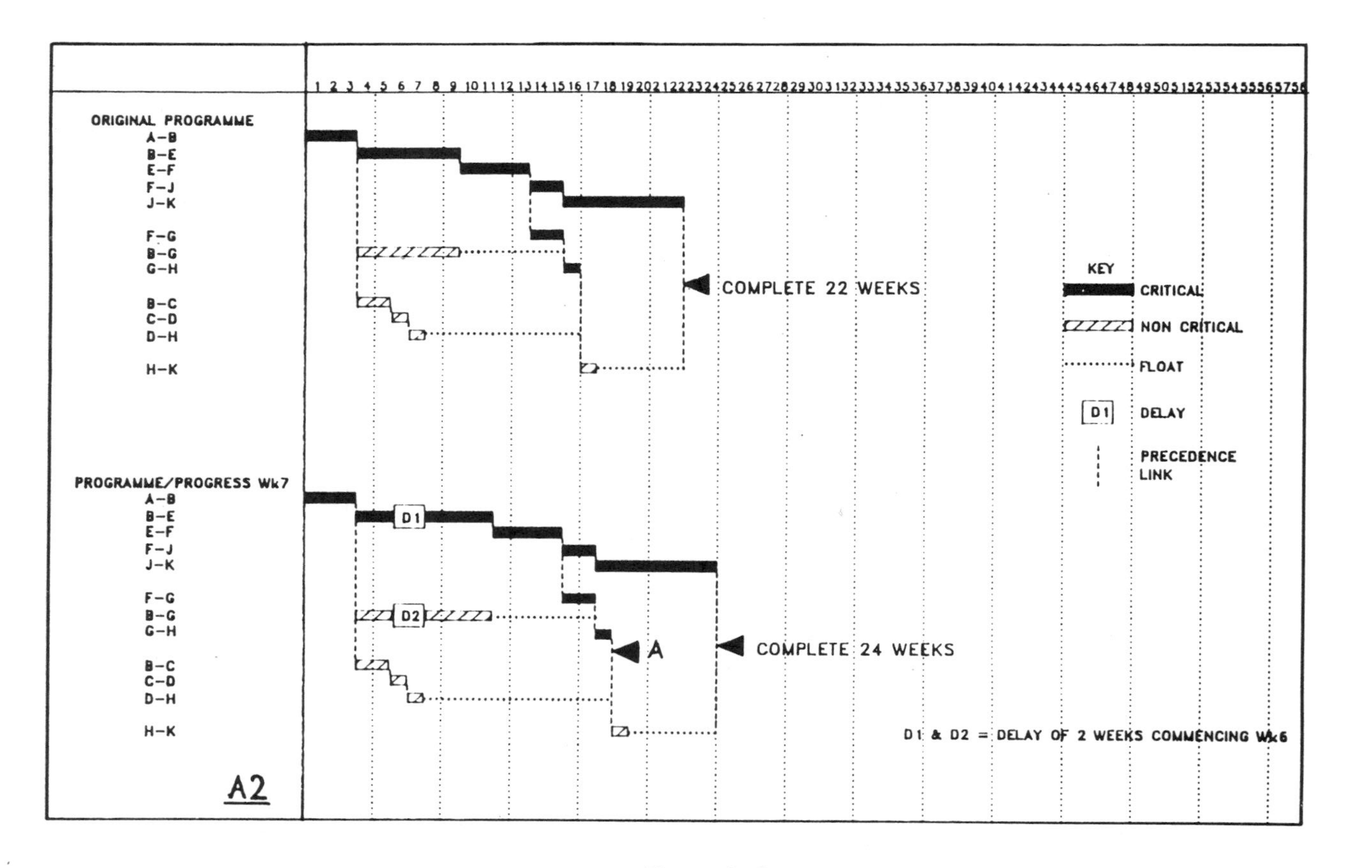

Figure A. 2

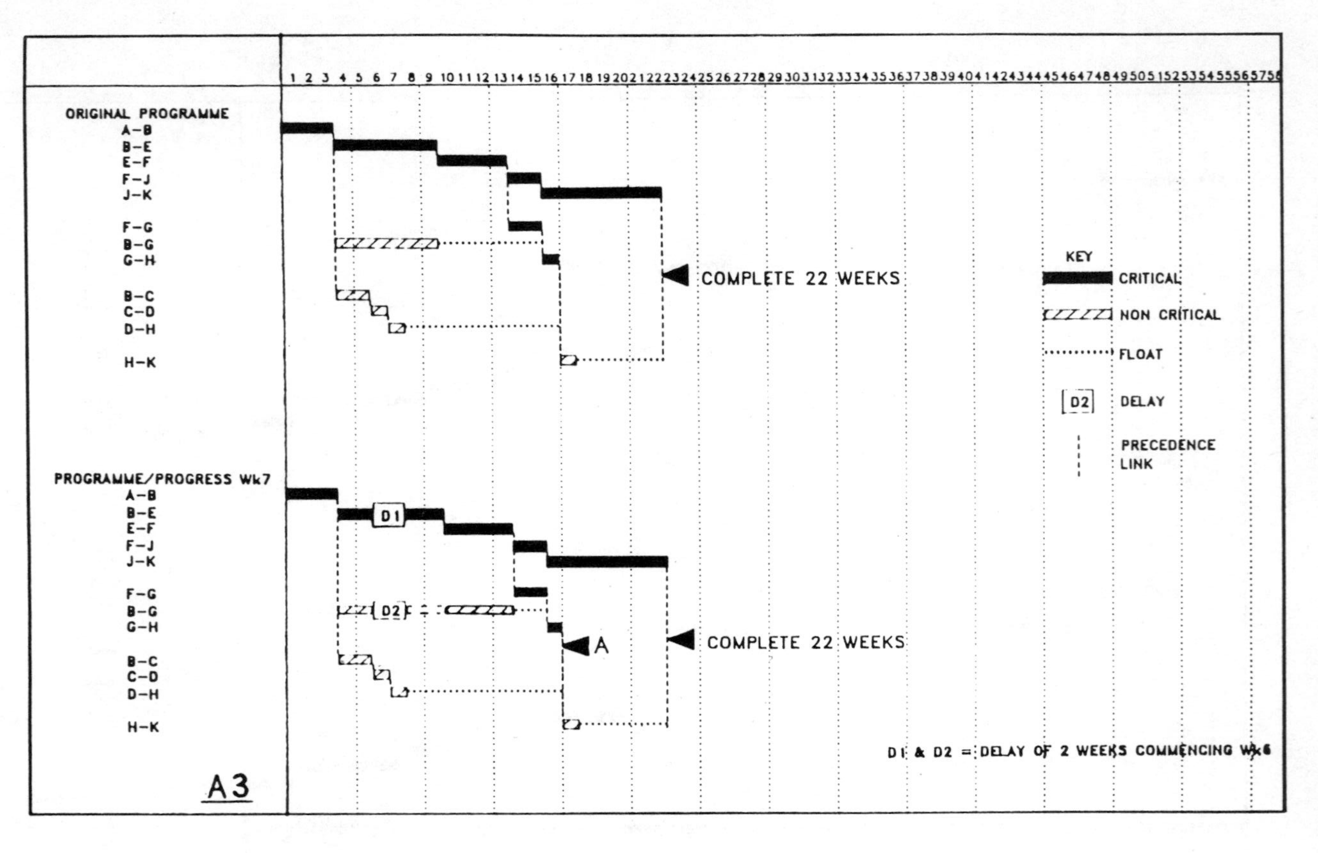

1 2 3 4 5 6 7 8 9 10 11 12 13 14 15 16 17 18 19 20 21 22 23 24 25 26 27 28 29 30 31 32 33 34 35 36 37 38 39 40 41 42 43 44 45 46 47 48 49 50 51 52 53 54 55 56 57
ORIGINAL PROGRAMME
A–B
B–E
E–F
F–J
J–K
F–G
B–G
G–H
B–C
C–D
D–H
H–K
COMPLETE 22 WEEKS
KEY
CRITICAL
NON CRITICAL
FLOAT
D2
DELAY
PRECEDENCE LINK
PROGRAMME/PROGRESS Wk7
A–B
B–E
E–F
F–J
J–K
F–G
B–G
G–H
B–C
C–D
D–H
H–K
D1
D2
A
COMPLETE 22 WEEKS
D1 & D2 = DELAY OF 2 WEEKS COMMENCING Wk6
A3

Figure A. 3

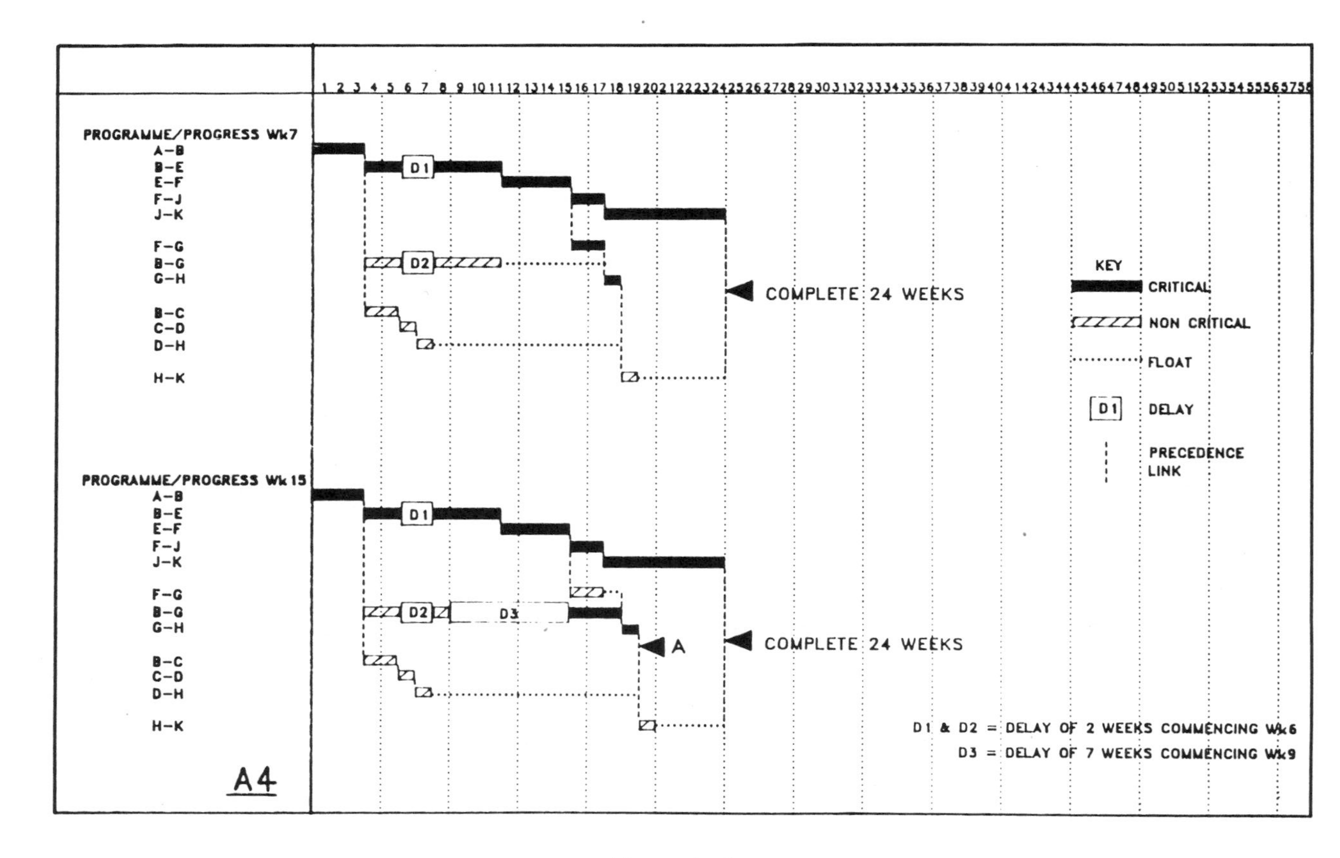
PROGRAMME/PROGRESS Wk7
A–B
B–E
E–F
F–J
J–K
F–G
B–G
G–H
B–C
C–D
D–H
H–K
D1
D2
COMPLETE 24 WEEKS
PROGRAMME/PROGRESS Wk15
D3
A
KEY
CRITICAL
NON CRITICAL
FLOAT
DELAY
PRECEDENCE LINK
D1 & D2 = DELAY OF 2 WEEKS COMMENCING Wk6
D3 = DELAY OF 7 WEEKS COMMENCING Wk9
A4

Figure A. 4

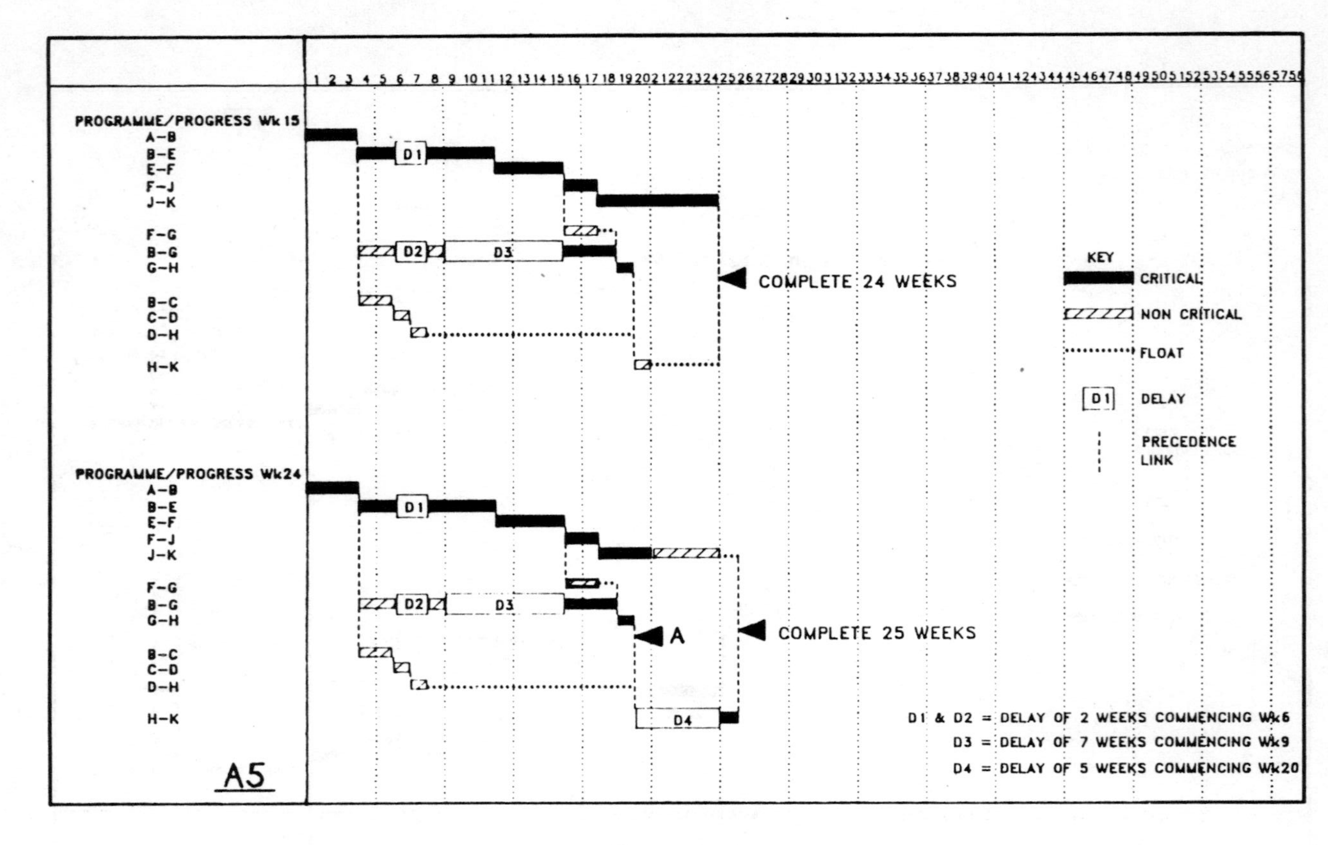
PROGRAMME/PROGRESS Wk 15
A–B
B–E
E–F
F–J
J–K
F–G
B–G
G–H
B–C
C–D
D–H
H–K
D1
D2
D3
COMPLETE 24 WEEKS
KEY
CRITICAL
NON CRITICAL
FLOAT
DELAY
PRECEDENCE LINK
PROGRAMME/PROGRESS Wk24
A
COMPLETE 25 WEEKS
D4
D1 & D2 = DELAY OF 2 WEEKS COMMENCING Wk6
D3 = DELAY OF 7 WEEKS COMMENCING Wk9
D4 = DELAY OF 5 WEEKS COMMENCING Wk20
A5

Figure A. 5

Appendix III

BETTER BUILDERS LTD
FINANCE CHARGES ON BALANCE DUE

DATE	CAPITAL ADDED £	CAPITAL TOTAL £	RATE	PERIOD DAYS	INTEREST £	CAP + INT £
01.05.91	27256.00	27256.00	0.140	24.00	250.90	
24.05.91		27256.00	0.135	7.00	70.57	
01.06.91		27256.00	0.135	30.00	302.43	27879.90
01.07.91*		27879.90	0.135	11.00	113.43	
12.07.91		27879.90	0.130	20.00	198.60	
01.08.91	17720.01	45599.91	0.130	31.00	503.47	
01.09.91	63052.43	108652.34	0.130	4.00	154.79	
04.09.91		108652.34	0.130	26.00	1006.15	110628.78
01.10.91*		110628.78	0.125	31.00	1174.48	
01.11.91		110628.78	0.125	61.00	2311.49	114114.35
01.01.92*		114114.35	0.125	31.00	1211.49	
01.02.92	100.41	114214.76	0.125	29.00	1131.23	
01.03.92	50.28	114265.04	0.125	31.00	1209.77	117817.52
01.04.92*		117817.52	0.125			
TOTAL & CHECK	108179.13			336.00	9638.39	117817.52

* Rest days for compounding interest
A:BB1

Architect's reply to the contractor's letter of 31 March 1992 and the claim submission:

Date 6 May 1992

Dear Sirs,

Re: ABC Stores and Depot, New Road, Lower Hamstead, Wilton.

I refer to your letter and enclosures of 31 March 1992.

Extensions of time
Having considered the arguments in your submission, I am prepared to fix later completion dates of 23 June 1991 for section A and 4 August 1991 for the works. That is, total extensions of time of three weeks inclusive of the extensions already made in my certificate EOT 1 dated 12 August 1991. I am not empowered to deal with the matter of finance charges on liquidated damages, and I am instructed to inform you that the employer wishes to discuss this with you at a meeting to be arranged next week. In the meantime, I will prepare the necessary certificate and issue it by the end of this week.

Loss and/or expense
I cannot agree that you are entitled to prolongation costs for the period of prolongation caused by delays (D2) and (D3). The principal cause of delay during this period was exceptionally adverse weather conditions (delay D1). I have considered your arguments on reprogramming (paragraph 4.4.1 of your submission) and I reject it on the grounds that you would have required additional formwork to make any progress on activity B-E in order to mitigate the delay. No additional formwork was delivered to site for this work.

Further, I cannot agree that your resources were prevented from taking on other work as a result of the delay (D4). According to my records, site offices were removed in week 24 and your resources were decreased commencing the end of week 23. I am prepared to include the part-time cost of your general foreman as part of your claim (subject to substantiation of his time spent on site). I do not accept that you lost any opportunity to make a contribution to overheads and profit as a result of one week delay. Even if I allowed loss of overheads and profit for any part of the prolonged period, I would have to deduct the overheads and profit recovered in the variations and extra work to activity B-G.

I also reject your argument on reimbursement of the costs of preparing the claim.

The quantity surveyor's assessment of loss and/or expense, taking into account the above comments, is £18 500.00 inclusive of finance charges up to the date of this letter.

A statement pursuant to clause 30.6.1 of the conditions of contract will be sent to you within the next few weeks.

Yours faithfully

T. Square

Contractor's reply to the architect's letter of 6 May 1992:

Date 14 May 1992

Dear Sir

Re: ABC Stores and Depot, New Road, Lower Hamstead, Wilton.

Thank you for your letter of 6 May 1992.

We cannot agree with your comments on our claim for loss and/or expense and/or damages.

Regarding measures to mitigate the delay caused by exceptionally adverse weather conditions (delay D1), the work which would have been done in the first week after the delay [week 8] was the excavation of a trench 2.5 metres wide by 2.25 metres deep. No formwork was required until the second week. We enclose herewith the acknowledgement of order for additional formwork which was due to be delivered on 6 April 1991. Accordingly, had we carried out the measures to mitigate the delay, we would have been able to complete activity B-E in accordance with our original programme.

Regarding the removal of site offices and reduction in resources, we had originally planned to remove the site offices before the completion date and our resources would have been reduced commencing week 20 if the project had not been delayed. As a result of delays (D2), (D3) and (D4) our resources were required for this project for three weeks longer than they would have been if there had been no delay. We reject the argument that we did not lose any opportunity to make a contribution to overheads and profit as a result of the delay. Please find enclosed a copy of the minutes of our board meeting on 1 July 1991 in which it it is recorded that we postpone commencement of our own speculative development of twenty-six houses because our labour, staff and plant were retained on this project as a result of the delay.

We also disagree with the proposition that an adjustment should be made for overheads and profit recovered in variations and extra work. This work delayed activity B-G and delayed completion of section A. There was no effect on the period of prolongation (which was a result of late nomination of Pumps & Co). In other words, the overheads and profit recovered in the additional work to activity B-G would have been earned within the original contract period and no adjustment would be have been made (see *The Presentation and Settlement of Contractors' Claims* by Geoffrey Trickey at pp 127 and 128).

In the circumstances of this case, we must insist that it is right to reimburse the cost of preparing the claim.

We trust that you will reconsider the matter at your earliest convenience.

Yours faithfully

For and on behalf of Better Builders Ltd.

Footnotes

Negotiations are in progress. Some of the arguments in the above example may be persuasive in negotiations. Differences of opinion in the industry on the use of a formula, concurrent delays, adjustment for overheads and profit recovered in variations and the costs of preparing the claim may give rise to real stumbling blocks in the negotiations to settle the sums in dispute.

This example may not cover all that went wrong during the progress of the works. There may have been other delays by the contractor. However, on the facts described in the example, the contractor appears to have reasonable grounds to pursue his claims.

While, in this case, the architect has now granted an extension for the full period of delay, some practitioners may argue that the words used in clause 25.3.1 of JCT80:

> 'If, in the opinion of the Architect,... any of the events... are a Relevant Event *and the completion of the Works is likely to be delayed thereby*

> *beyond the Completion Date*... the Architect shall in writing... give an extension of time...'

do not cover extensions of time in the circumstances of this case. For example, none of the delays (D2), (D3) or (D4) caused completion of the works (or section A) to be delayed beyond the completion date. Delay (D1) had already caused the completion of the works and section A to be delayed (or likely to be delayed) beyond the completion date. Unless clause 25.3.3 is intended to allow greater flexibility for granting extensions of time, it would appear to be at least arguable that once the contractor has caused delay which was likely to cause completion of the works to be delayed beyond the completion date, the clause does not bite. If that was the case, there would be no valid extension of time provision (after the contractor's delay) and all subsequent delays within the control of the employer would put time at large and no liquidated damages could be recovered. This is clearly not the intention of the contract, but some revised drafting may be helpful. Clause 23 of JCT63 (which is still in use in some parts of the world) does not have any provisions similar to clause 25.3.3 of JCT80, in which case the clause may be defective if construed very narrowly.

References

Books and publications

Hudson's Building and Engineering Contracts, Tenth Edition
Author: I. N. Duncan Wallace
Publisher: Sweet & Maxwell Limited, 1970

Hudson's Building and Engineering Contracts, Tenth Edition, First Supplement
Author: I. N. Duncan Wallace
Publisher: Sweet & Maxwell Limited, 1978

A Building Contract Casebook, Second Edition
Authors: Dr Vincent Powell-Smith and Michael Furmston
Publisher: BSP Professional, 1990

The Railway Navvies
Author: Terry Coleman
Publisher: Penguin Books, Reprinted 1981

House and Cottage Construction, Volume IV
Author: Horace W. Langdon (Chapter II)
Publisher: Caxton Publishing Company Limited, Estimated 1927

Engineering Law and the I.C.E. Contracts
Author: Max W. Abrahamson
Publisher: Elsevier Applied Science Publishers, 1979

Emden's Building Contracts and Practice, Eighth Edition, Volume 2
Author: S. Bickford-Smith
Publisher: Butterworths, 1990

Construction Contracts: Principles and Policies in Tort and Contract
Author: I. N. Duncan Wallace
Publisher: Sweet & Maxwell Ltd, 1986

Keating on Building Contracts, First Supplement to the Fourth Edition
Author: Donald Keating
Publisher: Sweet & Maxwell Ltd, 1982

Keating on Building Contracts, Fifth Edition
Author: Hon Sir Anthony May, M.A.
Publisher: Sweet & Maxwell Ltd, 1991

The Presentation and Settlement of Contractors' Claims
Author: Geoffrey Trickey
Publisher: E. & F.N. Spon Ltd, 1983

Building Law Information Subscriber Service (BLISS)
Weekly Bulletins: BLISS Annuals
Authors: Roger Knowles and Mark Entwhistle, Edited by Ann Glacki
Publisher: Knowles Publications, 1988, 1989, 1990, 1991

Claims: Their Mysteries Unravelled, First and Second Editions
Authors: Roger Knowles and David Carrick
Publisher: Knowles Publications, 1989

Recent Legal Cases
Author: Roger Knowles
Publisher: Knowles Publications, 1991

Thirty Crucial Contractual Issues and Their Solutions
Author: Roger Knowles
Publisher: Knowles Publications, 1992

Alternative Dispute Resolution
Author: Patrick O'Connor
Publisher: Knowles Publications, 1991

The Banwell Report – The Placing and Management of Contracts for Building and Civil Engineering Work, (HMSO), 1964

The Public Supplies Directive, 77/62/EEC, amended 22 March 1988 88/295/EEC (EC)

The Public Works Directive, 77/305/EEC, amended 18 July 1989 89/440/EEC (EC)

The Excluded Sectors Directive, 90/531 OJ L2971/29 October 1990 (EC)

International Construction, November 1980: *Anatomy of a Construction Project*
Author: Kris Nielson

Public Procurement Directives
Author: Robert Falkner
Conference paper at seminar '*The Construction Industry, Europe and 1992*', organised by Legal Studies & Services Limited, 10 December 1990

The Arbitration Acts of 1950, 1975 and 1979

The Courts and Legal Services Act 1990

Practice Note 20, issued by the Joint Contracts Tribunal, 1988

List of cases

J. and J.C. Abrahams v. *Ancliffe* [1938] 2 NZLR 420

Alghussein Establishment v. *Eton College* [1988] 1 WLR 587

Amalgamated Building Contractors v. *Waltham Holy Cross UDC* [1952] 2 All ER 452

Appeal of Eichleay Corporation, ASBCA 5183, 60–2 BCA (CCH) 2688 (1960)

Baese Pty Ltd v. *R.A. Bracken Building Pty Ltd* (1989) 52 BLR 130

Bolt v. *Thomas* (1859): *(Hudson's Building and Engineering Contracts, Tenth Edition* at page 196)

Boyd & Forrest v. *Glasgow S.W. Railway Company* [1914] SC 472

Bramall and Ogden v *Sheffield City Council* (1983) 29 BLR 73

Bremer Handelsgesell-Schaft M.B.H. v. *Vanden Avenne-Izigem P.V.B.A* [1978] 2 Lloyds LR 109

British Steel Corporation v. *Cleveland Bridge Engineering Co Ltd.* (1981) 24 BLR 94

Bryant and Sons Ltd v. *Birmingham Saturday Hospital Fund* [1938] 1 All ER 503

Bush v. *Whitehaven Port and Town Trustees* (1888) 52 JP 392

Capital Electric Company v. *United States* (Appeal No 88/965, 7.2.84) 729 F. 2d 143 (1984)

Carr v. *J.A. Berriman Pty Ltd* (1953) 27 ALJR 237

Carslogie S.S. Co. v. *Norwegian Government* [1952] AC 292

C.J.Sims v. *Shaftesbury Plc* (1991) QBD; 8-CLD-03–10

Commission of the European Communities v. *Ireland* (1988) 44 BLR 1

J. Crosby & Sons Ltd v. *Portland Urban District Council* (1967) 5 BLR 121

Davis Contractors Limited v. *Fareham U.D.C.* [1956] AC 696

Department of Environment for Northern Ireland v. *Farrans* (1981) 19 BLR 1

Ellis-Don v. *Parking Authority of Toronto* (1978) 28 BLR 98

Fairclough Building Ltd v. *Rhuddlan Borough Council* (1985) 30 BLR 26

H. Fairweather & Co Ltd v. *London Borough of Wandsworth* (1987) 39 BLR 106

A.E. Farr Ltd v. *Ministry of Transport* (1965) 5 BLR 94

Fernbrook Trading Co. Ltd v. *Taggart* [1979] 1 NZLR 556

J.F. Finnegan v. *Sheffield City Council* (1989) 43 BLR 124

Fratelli Costanzospa SpA v. *Comune di Millano* [1990] 3 CMLR 239

Gilbert Ash (Northern) Ltd v. *Modern Engineering (Bristol) Ltd* [1974] AC 689

GKN Centrax Gears Ltd v. *Malbro Ltd.* [1965] 2 Lloyds LR 555

M.J. Gleeson (Contractors) Ltd v. *London Borough of Hillingdon* (1970) 215 EG 165

Glenlion Construction Ltd v. *The Guinness Trust* (1987) 39 BLR 89

Government of Ceylon v. *Chandris* [1965] 3 All ER 48

Holme v. *Guppy* (1838) 3 M & W 378

Hong Kong Teakwood Limited v. *Shui On Construction Company Limited* (1984) HKLR 235

Howard Marine & Dredging v. *Ogden* (1978) 9 BLR 34

James Longley & Co Ltd v. *South West Regional Health Authority* (1985) 25 BLR 56

Leyland Shipping Company v. *Norwich Union Fire Insurance Society* [1918] AC 350

London Borough of Merton v. *Stanley Hugh Leach Ltd* (1985) 32 BLR 51

Marsden Construction Co Ltd v. *Kigass Ltd* (1989) 15 ConLR 116

Martin Grant & Co Ltd v. *Sir Lindsay Parkinson & Co Ltd* (1984) 29 BLR 31

Mathind Ltd v. *E. Turner & Sons Ltd* [1986] 23 ConLR 16

McMaster University v. *Wilchar Construction Ltd* (1971) 22 DLR (3d) 9

Michael I. Warde v. *Feedex International, Inc* [1984] 1 Lloyds LR 310

Michael Salliss & Co Ltd v. *E.C.A. Calil and William F. Newman & Associates* [1989] 13 Con LR 68

Mid-Glamorgan County Council v. *J. Devonald Williams & Partner* (1991), Unreported

Miller v. *London County Council* (1934) 151 LT 425

Mitsui Construction Co Ltd v. *Attorney General of Hong Kong* (1986) 33 BLR 1

Moon v. *Witney Union* (1837): (*Hudson's Building and Engineering Contracts, Tenth edition* at page 113)

Morgan Grenfell v. *Sunderland Borough Council and Seven Seas Dredging Ltd* (1991) 51 BLR 85

Morrison-Knudsen International Co Inc and Another v. *Commonwealth of Australia* (1980) 13 BLR 114

Morrison-Knudsen v. *B.C. Hydro & Power* (1975) 85 DLR 3d 186

Nash Dredging Ltd v. *Kestrell Marine Ltd* (1986) SLT 62

North West Regional Health Authority v. *Derek Crouch* [1984] 2 WLR 676

Pacific Associates Inc and Another v. *Baxter and Others* (1988) 44 BLR 33

Peak Construction (Liverpool) Ltd v. *Mckinney Foundations Ltd* (1970) 1 BLR 111

Penvidic Contracting Co. Ltd v. *International Nickel Co. of Canada Ltd* (1975) 53 DLR (3d) 748

Percy Bilton Ltd v. *The Greater London Council* (1981) 17 BLR 1 (CA), (1982) 20 BLR 1 (HL)

Perini Pacific Ltd v. *Greater Vancouver Sewerage and Drainage District Council* [1967] SCR 189

Rapid Building Group Ltd v. *Ealing Family Housing Association Ltd* (1984) 29 BLR 5

Rees and Kirby Ltd v. *Swansea City Council* (1985) 30 BLR 1

Schindler Lifts (H.K) Ltd v. *Shui On Construction Company Limited* (1984) 29 BLR 95

Sutcliffe v. *Thackrah and Others* (1974) 4 BLR 16

Tate & Lyle Food Distribution Ltd and Another v. *Greater London Council* [1982] 1 WLR 149

Temloc Ltd v. *Erril Properties Ltd* (1987) 39 BLR 31

Tersons Ltd v. *Stevenage Development Corporation* (1963) 5 BLR 54

Token Construction Co Ltd v. *Charlton Estates Ltd* (1976) 1 BLR 48

Tramountana Armadora SA v. *Atlantic Shipping Co., SA* [1978] 2 All ER 870

Waghorn v. *Wimbledon Local Board* (1877): (*Hudson's Building and Engineering Contracts, Tenth edition* at page 114)

Wates Construction (London) Ltd v. *Franthom Properties Ltd* (1991) 53 BLR 23

Wegan Construction Company Pty. Ltd. v. *Wodonga Sewerage Authority* [1978] VR 67

Wells v. *Army and Navy Co-operative Society Ltd* (1902) 86 LT 764

Wharf Properties Ltd and Another v. *Eric Cumine Associates, and Others* (1988) 45 BLR 72 (1991) 52 BLR 52 P.C.

Whittall Builders Company Ltd v. *Chester-le-Street District Council* (1985) Unreported

William Lacey (Hounslow) Ltd v. *Davis* [1987] 2 All ER 712

Yorkshire Water Authority v. *Sir Alfred McAlpine and Son (Northern) Ltd* (1985) 32 BLR 114

Abbreviations used in case references

AC: *Law Reports Appeal Cases*

ALJR: *Australian Law Journal Reports*

All ER: *All England Law Reports*

BCA: *Board of Contract Appeals (USA)*

BLR: *Building Law Reports*

CA: *Court of Appeal (UK)*

CLD: *Construction Law Digest*

CMLR: *Common Market Law Reports*

ConLR: *Construction Law Reports*

DLR: *Dominion Law Reports*

EG: *Estates Gazette*

F: *Federal Circular (USA)*

HKLR: *Hong Kong Law Reports*

HL: *House of Lords*

JP: *Justice of the Peace and Local Government Review*

Lloyds LR: *Lloyds Law Reports*

LT: *Law Times Report*

M & W: *Meeson & Welsby*

NZLR: *New Zealand Law Reports*

PC: *Privy Council*

QBD: *Queens Bench Division*

SC: *Session Cases*

SCR: *Supreme Court Reports*

SLT: *Scots Law Times*

VR: *Victorian Reports*

WLR: *Weekly Law Reports*

Forms of contract

JCT63: *The Standard Form of Building Contract* issued by the Joint Contracts Tribunal in 1963

JCT80: *The Standard Form of Building Contract* issued by the Joint Contracts Tribunal in 1980

IFC84: *The Intermediate Form of Building Contract* issued by the Joint Contracts Tribunal in 1984

GC/Works/1: *General Conditions of Government Contracts for Building and Civil Engineering Works* prepared by the Department of the Environment (HMSO)

ICE Conditions of Contract: *The Conditions of Contract and Form of Tender, Agreement and Bond for use in connection with Works of Civil Engineering Construction* issued by The Institution of Civil Engineers, The Association of Consulting Engineers and The Federation of Civil Engineering Contractors

The Standard Form of Building Contract with Contractor's Design 1981 Edition, issued by the Joint Contracts Tribunal

CDPS: *Contractor's Designed Portion Supplement*, issued by the Joint Contracts Tribunal

MW80: *Agreement for Minor Building Works,* issued by the Joint Contracts Tribunal

The Standard Form of Contract with Approximate Quantities, issued by the Joint Contracts Tribunal

The Fixed Fee Form of Contract, issued by the Joint Contracts Tribunal

The Standard Form of Management Contract, issued by the Joint Contracts Tribunal

NEC: *New Engineering Contract* (1991), issued by the Institution of Civil Engineers

FIDIC Conditions of Contract: *The Conditions of Contract* issued by The Federation Internationale des Engenieurs-Conseils

SIA Conditions of Contract: *The Conditions of Contract* issued by the Singapore Institute of Architects

RIBA Model Form of Contract issued by the Royal Institute of British Architects

NSC/1: *The Standard Form of Nominated Subcontract Tender and Agreement,* issued by the Joint Contracts Tribunal

NSC/4A: *Standard Form of Nominated Subcontract,* issued by the Joint Contracts Tribunal

Miscellaneous abbreviations

BAS: Building Automation System

BLISS: Building Law Information Subscriber Service

CCTV: Closed Circuit Television

CESMM: *Civil Engineering Standard Method of Measurement,* issued by the Institution of Civil Engineers

EC: European Commission

EEC: European Economic Community

ECU: European Currency Unit

FIDIC: Federation Internationale des Engenieurs-Conseils

HMSO: Her Majesty's Stationery Office

HVAC: Heating Ventilating and Air Conditioning

ICE: Institution of Civil Engineers

ISO: International Standards Organisation

JCT: Joint Contracts Tribunal

OJ: The Official Journal of The European Commission

PC: Prime Cost

RIBA: Royal Institute of British Architects

SIA: Singapore Institute of Architects

SMM: *Standard Method of Measurement*: Published by the Royal Institution of Chartered Surveyors and The National Federation of Building Trades Employers (Building Employers Confederation)

SMM6: *Standard Method of Measurement (Sixth Edition)* issued by the Royal Institution of Chartered Surveyors and the National Federation of Building Trades Employers (Building Employers Confederation)

Index